ENCYCLOPAEDIA OF
PSYCHOLOGY

ENCYCLOPAEDIA OF
PSYCHOLOGY

Jagdish Chaudhary

ANMOL PUBLICATIONS PVT. LTD.
NEW DELHI - 110 002 (INDIA)

ANMOL PUBLICATIONS PVT. LTD.
H.O.: 4374/4B, Ansari Road, Darya Ganj,
New Delhi-110 002 (India)
Ph.: 23278000, 23261597

B.O.: No. 1015, Ist Main Road, BSK IIIrd Stage
IIIrd Phase, IIIrd Block
Bangalore - 560 085 (India)
Visit us at: www.anmolpublications.com

Encyclopaedia of Psychology

First Published, 2006
ISBN 81-261-3029-6

PRINTED IN INDIA

Printed at Mehra Offset Press, Delhi.

Preface

This volume covers all aspects of Psychology and its role in society. This Encyclopaedia covers a range of various Psychology disciplines such as clinical, developmental, educational, historical, occupational, personality, philosophical, physiological (human and animal), and social psychology. Importance is placed throughout on intersections with related disciplines such as sociology, neuroscience, and artificial intelligence. Schools of psychology, important practitioners, fundamental concepts, and empirical techniques are also treated in articles.

In recent years there have been dramatic changes in Psychology on both national scale and international scale. These changes have had a major impact on all our lives. Not only this field is becoming more complex, but even terminology is becoming complex day by day. Our aim of compiling this encyclopaedia of psychology is to provide a useful baseline set of key concepts for examining the difficult topics of laws, theories, principles, effects, doctrines, and hypotheses in psychology.

This new encyclopaedia integrates all of the diverse approaches that have gone into the making of the modern field of Psychology. It is unique among available reference works in that it covers the most recent topics of research. Encyclopaedia of Psychology is intended for all those interested in this field. It will be of immense value to students of all kinds, especially students of Psychology. It will also be a major source of reference for professionals and students, who require an authoritative and up-to-date guide to assist them in their work.

In preparation of this book, the author has freely consulted large number of books and journals so no authenticity is claimed. Author is especially thankful to Anmol Publications Pvt. Ltd., New Delhi for shaping this book in its final form. Suggestions for further improvement of this book are not only welcome but also greatly appreciated.

Author

A

AB design A single subject research design that contains one baseline (A) and one treatment (B).

A-B-A design Experimental design in which participants first experience the baseline condition (A), then experience the experimental treatment (B), and then return to the baseline (A).

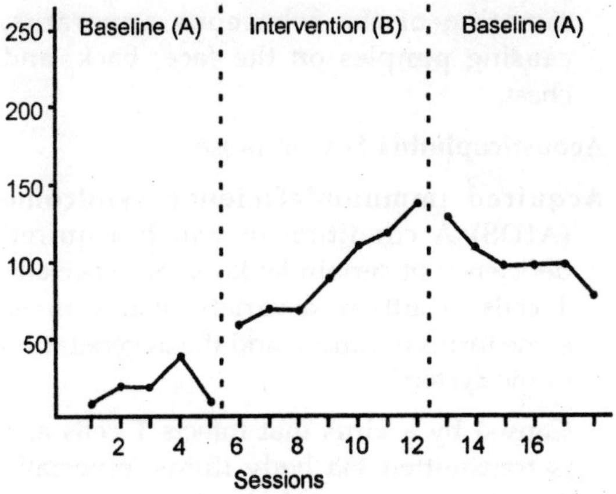

Fig. A-B-A design

ABAB design A single subject research design that contains a baseline (A1), treatment (B1), a second baseline (B2) and a second treatment phase (B2).

Abasement The need to comply, surrender, confess, or atone. A need to accept punishment.

Abduction In conditional reasoning, the generation of an explanation for an event, taken from a theory of how the world works, the plausibility of which depends in large part on the number and the likelihood of the alternative explanations.

Ablation The removal of tissue with a vacuum or scalpel, sometimes used as a research technique (allowing investigators to ask how an organism functions in the absence of this tissue).

Ablutophobia Fear of washing or bathing.

Abnormal psychology Abnormal psychology studies the nature of psychopathology, its causes, and its treatments. Of course the definition of what constitutes 'abnormal' has varied across time and across cultures. Individuals also vary in what they regard as normal or abnormal behaviour.

Absolute refractory period The period during which a neuron lies dormant after an action potential has been completed.

Absolute threshold The minimum amount of stimulation needed for a person to detect the stimulus 50 percent of the time.

Absolute zero Characteristic of a scale of measurement that contains a point where the scale has no value.

Abstract identifications A category in the analysis of responses to the 'who am I?', which includes self-descriptions that are too abstract or general to constitute social identity elements (e.g. A human being).

Acarophobia Fear of itching or of the insects that cause itching.

Acceleration programs Rapid promotion through advanced studies for students who are gifted or talented.

Accentuation theory One of a variety of social cognition theories, which suggests that when other people's attitudes differ from our own, we tend to exaggerate the difference between them.

Acceptance One of Elisabeth Kubler-Ross's stages of dying, in which the dying come to terms with their approaching death.

Accessory structures In sensory processing, the parts of sensory systems that gather external stimulus energies and fashion the proximal stimulus, which the receptors then transduce.

Accommodation Changes in the thickness of the lens of the eye that focus images of near or distant objects on the retina. Also, the process by which existing schemata are modified or changed by new experiences.

Accommodative distortion Retrospective alterations of memory to fit a schema.

Accountability The degree to which people are held responsible for their task performances or decision outcomes.

Acculturation The process of acquiring or adapting to a new culture, its traditions, customs, and patterns of daily living.

Acerophobia Fear of sourness.

Acetylcholine A neurotransmitter found in many parts of the nervous system. Among many other functions, it serves as an excitatory transmitter at the synaptic junctions between muscle fibres and motor neurons.

Achievement batteries Standardized tests that include several subtests designed to measure knowledge of particular subjects.

Achievement motivation The desire to experience success and to participate in activities in which success is dependent on personal effort and abilities.

Achievement motive An impulse to master challenges and reach a high standard of excellence.

Achievement tests An assessment that measures skills and knowledge that people have already learned.

Achluophobia Fear of darkness.

Achromatic colours Colours, such as black, white, and the neutral greys, that do not have the property of hue.

Acne vulgaris A common, chronic skin disease, especially among adolescents and young adults, characterized by inflammation of the sebaceous apparatus, causing pimples on the face, back, and chest.

Acousticophobia Fear of noise.

Acquired immunodeficiency syndrome (AIDS) A condition in which acquired deficiency of certain leukocytes, especially T cells, results in a variety of infections, some forms of cancer, and the degeneration of the system.

Caused by a virus that infects T cells and is transmitted via body fluids, especially sexual secretions and blood.

Acquisition behaviour Behaviour that results in the addition of a thing or event to a situation.

Acquisition The stage in a classical conditioning experiment during which the conditioned response is first elicited by the conditioned stimulus.

Acronym A word made out of the first letters of several words.

Acrophobia Fear of heights.

Across- fibre theory The theory that a certain sensory quality is signalled by the pattern of neural activity across a number of different nerve fibres.

Acrostic A sentence or phrase in which each word begins with a letter that acts as a memory cue.

Action potential As the travelling signals of nerves and as the localized changes that contract muscle cells, action potentials are an essential feature of animal life. They set the pace of thought and action, may constrain the sizes of evolving anatomies and enable centralized control and coordination of organs and tissues. Non-propagating action potentials occur also in some plants.

Activation Activity of the visceral brain responsible for the fight-or-flight response in the face of danger; possible basis for neuroticism.

Activation synthesis theory The idea that during sleep, an automatic activation system in the brain produces a series of random electrical discharges that the sleeper roughly ties together by creating a storyline. An opposing theory to the traditional Freudian idea that dreams are secrets, fears, and the like rising from the unconscious.

Activational effect The effect of a hormone on a physiological system that has already developed. If the effect involves the brain, it can influence behaviour. An example is facilitation of sexual arousal and performance.

Activation-synthesis hypothesis An account which holds that dreams may reflect the brain's aroused state during REM sleep, when the cerebral cortex is active but shut off from sensory input. This helps explain the content and often disjointed form of REM dreams.

Activation-synthesis theory A theory proposing that neurons in the brain activate randomly during REM sleep.

Active euthanasia The deliberate termination of life to eliminate pain.

Active listening A feature of client-centred therapy that involves empathetic listening, by which the therapist echoes, restates, and clarifies what the client says.

Active span tasks Tasks in which research participants are asked to remember materials while simultaneously working on some other task; such tasks are an effective means of measuring working memory's capacity.

Activity dependence A property of neuronal plasticity such that changes in a neuron's functioning will occur only if that neuron is active (i.e., firing) at the same time as another neuron.

Activity theory Sees a positive correlation between keeping active and aging well.

Actor observer effect The tendency to attribute our own behaviour to situational causes and the behaviour of others to personal factors.

Actor-observer Actor-observer difference in the study of the attribution process, the tendency for actors to attribute their behaviour to the situation and for observers to attribute the actor's behaviour to dispositions.

Actor-observer difference The difference in attributions made by actors who describe their own actions and observers who describe another person's. The former emphasizes external, situational causes; the latter, internal, dispositional factors.

Act-outcome representation A type of association hypothesized by Edward Tolman to be the product of instrumental learning; an organism that has acquired this sort of association has acquired the knowledge that a certain type of act leads to a particular outcome.

Actualizing the self An explanation of individual behaviour as being a striving to fulfil, where self-image is a key filter which can either be in harmony with the organism or in opposition.

Actus reus The actual criminal act. A legal term referring to the actions behind the offence that must successfully be proven by the prosecution so that the defendant may be found guilty.

Acuity The ability to distinguish between separate points projected on the retina. Acuity is greatest in the fovea, where the receptors are closely bunched together.

Acute stress A transient state of arousal with typically clear onset and offset patterns.

Acute stress disorder A reaction sometimes observed in individuals who have experienced a traumatic event that is characterized by recurrent nightmares and waking flashbacks of the traumatic event.

Adaptation An inherited characteristic that increases in a population because it provides a survival or reproductive advantage.

Adaptation-level phenomenon The tendency to adapt to a given level of stimulation and thus to notice and react to changes from that level.

Adaptive behaviours Behaviours that increase reproductive success.

Adaptive significance The effectiveness of behaviour in aiding organisms in adjusting to changing environmental conditions.

Adaptive value In evolutionary terms, the extent to which an attribute increases the likelihood of viable offspring.

Addiction The physical and mental state of dependence on a substance or stimulus, to the point where withdrawal symptoms occur whenever the substance is not present in the body.

Additive colour mixture Mixing colours by stimulating the eye with two or more sets of wavelengths simultaneously (e.g., by focusing filtered light from two projectors on the same spot).

Additive rule The sum of the probabilities of separate (mutually exclusive) events determine the probability that any event will occur.

Additive strategy The process of listing the attributes of each element of a decision, weighing them according to importance, adding them up, and determining which one is more appealing based on the result.

Adequate stimulus An electrical pulse above the threshold, or critical point, that induces an action potential in a neuron.

Adipose cells The cells within the body that provide long-term storage of energy resources, usually in the form of fatty acids that can be converted to glucose when needed.

Adolescence Ages 12 to 19.

Adolescent growth spurt A noticeable increase in height and weight during adolescence.

Adoption studies Studies in which researchers examine trait similarities between adopted children and their biological and adoptive parents to figure out whether that trait might be inherited.

Adrenal cortex The outer part of the adrenal glands, which secretes corticosteroids.

Adrenal medulla The inner core of the adrenal gland, which regulates the release of epinephrine (adrenaline) and norepinephrine into the bloodstream.

Adrenaline It causes quickening of the heart beat, strengthens the force of the heart's contraction, opens up the bronchioles in the lungs and has numerous other effects. The secretion of adrenaline by the adrenal is part of the "fight-or-flight" reaction that we have in response to being frightened.

Adrenocorticotropic hormone (ACTH) A hormone released by the pituitary gland that stimulates release of corticosteroids from the adrenal cortex.

Adult attachment interview A type of interview devised to probe adults' memories of their own childhood experiences and relationships with their parents, and from which four basic classifications of parenting styles have been defined.

Adult learners Students age 25 or older.

Advance organizers Activities and techniques that orient students to the material before reading or class presentations.

Adversarial model A dispute-resolution system in which the prosecution and defence present opposing sides of the story.

Advice Advice an ethical obligation on the part of trained psychologists to offer qualified help to participants in an observational study when evidence of mental of physical problems arise in relation to them.

Aeroacrophobia Fear of open high places.

Aerobic exercises Physical activity that expends considerable energy, increases blood flow and respiration and thereby stimulates and strengthens the heart and lungs and increase the body's efficient use of oxygen.

Aeronausiphobia Fear of vomiting secondary to airsickness.

Aerophobia Fear of drafts, air swallowing, or airborne noxious substances.

Aesthetics of place In environmental psychology, the physical appearance of particular behavioural settings which affect the behaviour of those living within them.

Affect In environmental psychology, an individual's emotional response to the environment, which is predicted by the naturalness of the place; its upkeep, openness, order; and historical significance.

Affection Aim-inhibited sexuality. We connect affectionately because we can't spend all our time having sex.

Affective disorders Any disorder in which the primary symptoms are associated with mood disturbances, such as extreme depression, excessive elation, or both.

Affective forecasting The process of predicting how one would feel in response to future emotional events.

Afferent nerves Bundles of axons that carry information from muscles and sense organs to the central nervous system.

Affordances In environmental psychology, the kinds of activities that a place permits or encourages in those living within them.

After effect of visual movement An effect observed after one stares at a steadily moving object for a while. If one now looks at a stationary object, it appears to be moving in the direction opposite to the movement observed initially.

Afterimage A colour we perceive after another colour is removed.

Agateophobia Fear of insanity.

Age 30 transition A stage in the novice phase of early adulthood; early adulthood ranging from ages 28 to 33.

Age clock The internal sense of timing of physical and social events that determines the various life stages through which adults pass.

Age of viability The point at which a foetus has some chance of surviving outside the mother if born prematurely.

Ageism Discrimination against people on the basis of age; specifically, discrimination against, and prejudicial stereotyping of, older people.

Aggression The act of initiating hostilities or invasion. The practice or habit of launching attacks. Hostile or destructive behaviour or actions. In psychology aggression encompasses many different types behaviour, some of which are not clearly related to each other. Consequently, aggression has been a difficult term to provide one concise definition for.

Aggression machine The apparatus used to measure physical aggression in a laboratory.

Aggressive An interpersonal style where only the immediate needs of the self are considered rather than the needs of others. (As opposed to passive or assertive).

Aggressive script A guide for behaviour and problem solving that is developed and stored in memory, and is characterized by aggression.

Agliophobia Fear of pain.

Agnosia A serious disturbance in the organization of sensory information produced by lesions in certain cortical areas. An example is visual agnosia in which the patient can see but often does not recognize what it is that he sees.

Agonists Drugs that enhance the activity of a neurotransmitter, often by increasing the amount of transmitter substance available (e.g., by blocking reuptake or by increasing the availability of precursors).

Agoraphobia A disorder involving anxiety about situations from which escape would be difficult or embarrassing or places where there might be no help if a panic attack occurred.

Agrammatism A language disturbance; difficulty in the production and comprehension of grammatical features, such as proper use of function words, word endings, and word order. Often seen in cases of Broca's aphasia.

Agraphobia Fear of sexual abuse.

Agreeableness Tendency to be a nice person.

Agrizoophobia Fear of wild animals.

Agyrophobia Fear of streets or crossing the street.

Aichmophobia Fear of needles or pointed objects.

AIDS Acronym for acquired immune deficiency syndrome, a syndrome caused by a virus that damages the immune system and weakens the body's ability to fight infection.

Ailurophobia Fear of cats.

Akashik records The notion that anything that has happened is stored as universal memory permanently which can be accessed through psychic methods.

Alarm call Special, genetically programmed cry that impels members of a given species to seek cover. A biological puzzle, since it suggests a form of altruism in which the individual giving the call appears to endanger her own survival.

Albuminurophobia Fear of kidney disease.

Alcoholism An addiction to ethanol, which is the psychoactive agent in alcohol.

Alektorophobia Fear of chickens.

Algophobia Fear of pain.

Algorithm In computer problem solving, a procedure in which all of the operations are specified step-by-step.

Algorithmic method A method for solving problems, which, in contrast to heuristics, will always find a solution, if there is one, but which may be very slow.

All or none law The rule that the size of the action potential is unaffected by increases in the intensity of stimulation beyond the threshold level.

Alleles Alternative forms of the same gene.

Alliumphobia Fear of garlic.

Allocated time Time during which students have the opportunity to learn.

Allodoxaphobia Fear of opinions.

All-or-none law A law that describes the fact that all action potentials have the same amplitude regardless of the stimulus that triggered them.

Alpha activity Rhythmical, medium-frequency activity of the electroencephalogram, usually indicating a state of quiet relaxation.

Alpha The set (and acceptable) probability that the null hypothesis is true. Psychologists typically set their *alpha* at .05. An alpha of less than .05 can be interpreted to mean that the differences are not due to sampling error.

Alpha male The most dominant male in an animal group's dominance hierarchy.

Alpha waves Fairly regular EEG waves, between eight to twelve per second, characteristic of a relaxed, waking state, usually with eyes closed.

Alternate-forms reliability The ability of a test to produce the same results when two different versions of it are given to the same group of people.

Alternative hypothesis In statistics, the hypothesis that the null hypothesis is false, that an obtained difference is so far from zero that one has to assume that the mean difference in the population is greater than zero and that the experimental condition has some effect.

Altophobia Fear of heights.

Altruism In human and animal behaviour, the tendency to act in ways that benefit other individuals more than oneself, sometimes even where there is a real cost involved to the giver. Alzheimer's disease the most common form of senile dementia, a disease which usually manifests itself in old age, but which can also occur prematurely, and is related to certain, possibly genetic, abnormalities and degeneration in brain tissue. Ambiguous figure a kind of visual illusion in which a picture can be seen in two different ways, where the percept usually flips between the two alternative interpretations. Amnesia loss of memory syndrome, characterised by intact lts and preserved procedural memory and which is caused by damage to one of two specific brain regions, the temporal lobes or the diencephalon. Amniotic sac a kind of water bed inside the female uterus in which the human embryo develops into a foetus and which cushions the embryo, keeping it at a constant temperature. Amygdala an almond-shaped group of cells in the temporal lobes of the brain, which is a major.

Altruistic Motivated by the desire to improve another's welfare.

Alzheimer's disease A degenerative brain disorder characterized by memory loss followed by increasing disorientation and culminating in total physical and mental helplessness and death. One of the major sites of the destruction is a pathway of acetylcholine-releasing cells leading from the base of the forebrain to the cortex and hippocampus.

Amacrine cells Cells that integrate information across the retina; rather than sending signals toward the brain, amacrine cells link bipolar cells to other bipolar cells and ganglion cells to other ganglion cells.

Amathophobia Fear of dust.

Amaxophobia Fear of riding in a car.

Ambidextrous Able to use both hands with equal ease.

Ambiguity A perceptual object that may have more than "one interpretation.

Ambiguity (sentence meaning) The case in which a sentence (i.e., one surface structure) has two meanings (i.e., two underlying structures). (For example, "These missionaries are ready to eat" overheard in a conversation between two cannibals.).

Ambiguous figure A kind of visual illusion in which a picture can be seen in two different ways, where the percept usually flips between the two alternative interpretations.

Ambiguous language Language that can be understood in several ways.

Ambivalent sexism A form of sexism characterized by attitudes about women that reflect both negative, resentful beliefs and feelings and affectionate, chivalrous, but potentially patronizing beliefs and feelings.

Ambulophobia Fear of walking.

American Sign Language (ASL) The manual-visual language system of deaf persons in America.

Amino acid The building blocks of proteins.

Amnesia Amnesia is a condition in which memory is disturbed. The causes of amnesia are organic or functional. Organic causes include damage to the brain, through trauma or disease, or use of certain (generally sedative) drugs. Functional causes are psychological factors, such as defence mechanisms. Hysterical post-traumatic amnesia is an example of this. Amnesia may also be spontaneous, in the case of transient global amnesia. This global type of amnesia is more common in middle-aged to elderly people, particularly males, and usually lasts less than 24 hours.

Amnesiphobia Fear of amnesia.

Amniocentesis test The surgical procedure of inserting a hollow needle through the abdominal wall into the uterus of a pregnant woman and extracting amniotic fluid, which may be analyzed to determine the sex of the developing foetus or the presence of disease, genetic defects, and so on.

Amnion The innermost membrane of the sac enclosing the embryo; it is filled with a watery fluid called amniotic fluid.

Amniotic fluid The watery fluid that fills the amniotic sac and cushions the developing foetus against injury and shock and provides constant temperature in the amniotic sac.

Amniotic sac A kind of water bed inside the female uterus in which the human embryo develops into a foetus and which cushions the embryo, keeping it at a constant temperature.

Amphetamine psychosis A pattern of symptoms similar to those observed in paranoid schizophrenia, but produced by frequent and large doses of amphetamines.

Amphetamines Drugs that increase the availability of dopamine norepinephrine, causing increased arousal and excitement. Large doses may lead to frenetic hyperactivity and delusions.

Amplitude The height of a wave crest, often used as a measure of intensity of a sound or light wave.

Amychophobia Fear of scratches or being scratched.

Amygdala Located in the brain's medial temporal lobe, the almond-shaped amygdala (in Latin, corpus amygda-loideum) is believed to play a key role in the emotions. It forms part of the limbic system. In humans and other animals, it is linked to both fear responses and pleasure. Its size is positively correlated with aggressive behaviour across species. In humans it is the most sexually dimorphic brain structure, and shrinks by more than 30% in males upon castration. Conditions such as anxiety, autism, depression, narcolepsy, post-traumatic stress disorder, phobias and schizophrenia are suspected of being linked to abnormal functioning of the amygdala owing to damage, developmental problems, or neurotransmitter imbalance.

Anablephobia Fear of looking up.

Anaclitic The kind of love that is directed toward an object, usually a parent. Contrast narcissistic.

Anal character According to Freud, a personality type that derives from serious conflicts during the anal stage and is distinguished by three symptomatic traits: compulsive orderliness, stubbornness, and stinginess.

Anal character One fixed at the anal level of psychosexual development, when the libido charges the anus with energy. People stuck at this early stage are regarded as parsimonious, obstinate, hoarding, and perfectionistic.

Anal eroticism Erotic/sexual pleasure from activities associated with stimulation of the anal region. Studies have linked such eroticism to aggression towards the individual being penetrated.

Anal expulsive personality Stemming from the Anal stage, a child who becomes fixated due to over control transfers his or her unresolved anal (or control) issues into characteristics such as cruelty, pushiness, messiness, or disorganization.

Anal retentive personality Stemming from the Anal stage, a child who becomes fixated due to under control transfers his or her unresolved anal (or control) issues into characteristics such as compulsivity, stinginess, cleanliness, organization, and obstinance.

Anal stage Freud's second stage of psychosexual development where the primary sexual focus is on the elimination or holding onto feces. The stage is often thought of as representing a child's ability to control his or her own world.

Analgesic A pain reliever.

Analogical mapping In problem solving, the use of the solution of another problem as a pattern, by establishing a mapping between the domains from which the problems are drawn - the source (or base) domain - and the target domain.

Analogical reasoning In the study of intelligence, a form of problem solving involving the comparison of similarities between pairs, often in the form of pictures.

Analogical representation A representation that shares some of the physical characteristics of an object; for example, a picture of a mouse is an analogical representation because it looks like the small rodent it represents.

Analogical storing In environmental cognition, a method for storing spatial information as an internal spatial representation - i.e. In the head.

Analogies Relating new concepts to information students already understand.

Analysis by synthesis An extreme form of top-down processing, most often applied to speech perception, in which the listener is thought to make sense of the speech waveform by attempting mentally to produce the same sounds as the speaker.

Analysis of variance In statistics, analysis of variance (ANOVA) is a collection of statistical models and their associated procedures which compare means by splitting the overall observed variance into different parts. The initial techniques of the analysis of variance were pioneered by the statistician and geneticist Ronald Fisher in the 1920s and 1930s, and is sometimes known as Fisher's ANOVA or Fisher's analysis of variance.

Analysis, the fundamental rule of To say whatever what comes into one's mind without censoring any of it.

Analysis, the purpose of To give the ego more control over the repressed id impulses. "Where there is id, there shall ego be.".

Analytic intelligence According to some investigators, the type of intelligence typically measured by intelligence tests and crucial for success in academic pursuits.

Analytic psychology A branch of psychology that views the person as a constellation of compensatory internal forces in a dynamic balance.

Anatomical coding A means of representing information by the nervous system; different features are coded by the activity of different neurons.

Anchoring and adjustment One of three types of heuristic which is based on the idea that people often make estimates, of probabilities among other things, by taking an initial value, or anchor, and adjusting it.

Anchoring heuristic An insufficient adjustment up or down from an original starting value when judging the probable value of some event or outcome.

Ancraophobia Fear of wind. (Anemophobia).

Androgen Any male sex hormone (e.g., testosterone).

Androgen insensitivity syndrome An inherited condition caused by a lack of functioning androgen receptors. Because androgens cannot exert their effects, a person with XY sex chromosomes develops as a female, with female external genitalia.

Androgyny Possessing many traditionally masculine and feminine personality traits.

Androphobia Fear of men.

Anemophobia Fear of air drafts or wind. (Ancraophobia).

Anger An emotional reaction elicited by a number of unique stimulus, including restraint, aggression, threat, attack, and frustration. Anger is characterized by a strong autonomic nervous system response, particularly the sympathetic component.

Anginophobia Fear of angina, choking or narrowness.

Angiotensin A substance produced by the kidneys when there is a decrease in the amount of liquid passing through them, activating receptors in the brain that monitor the volume of blood and other fluids in the body.

Anglophobia Fear of England or English culture, etc.

Angrophobia Fear of anger or of becoming angry.

Animal cognition The cognitive capabilities of nonhuman animals; researchers trace the development of cognitive capabilities across species and the continuity of capabilities from nonhuman to human animals.

Animism The belief that all animals and all moving objects possess spirits providing their motive force.

Ankylophobia Fear of immobility of a joint.

Anomia A difficulty in finding words that is often experienced by people with brain injuries.

Anonymity In the study of crowds, the lack of identification of the individual which leads them to act in a less social or responsible way than they otherwise would.

Anorexia nervosa A disorder characterized by refusal to maintain a body weight in the normal range, intense fear about gaining weight, and highly distorted body image.

A-not-B effect The tendency of infants around nine months of age to search for a hidden object by reaching for place *A*, where it was previously hidden, rather than a new place *B*, where it was hidden most recently while the child was watching.

ANOVA Analysis of Variance.

Antagonists Drugs that impede the activity of a neurotransmitter, often by decreasing the amount available (e.g., by speeding reuptake and decreasing availability of precursors).

Antecedent events Conditions present prior to the occurrence of a behaviour.

Antecedent stimulus Event that comes before a behaviour.

Anterior/posterior Toward the front/back.

Anterograde amnesia A memory deficit suffered after some brain damage. It is an inability to learn and remember any information imparted after the injury, with little effect on memory for information acquired before the injury.

Anthrophobia or anthophobia Fear of flowers.

Anthropophobia Fear of people or society.

Anti cathexis A process of internal frustration that, evolved from experiences of external frustration, sets up and maintains a repression. It involves the withdrawal of libido from whatever is being repressed. Anti-cathexes also oppose the impulsive cathexes otherwise made by the id toward objects.

Antianxiety drug A "tranquilizer," which reduces anxiety. The most common include chlordiazepoxide (Librium) and diazepam (Valium).

Antibodies Proteins in the immune system that recognize antigens and help kill invading microorganisms.

Anticipatory anxiety A fear of having a panic attack; may lead to the development of agoraphobia.

Anticipatory coping Efforts made in advance of a potentially stressful event to overcome, reduce, or tolerate the imbalance between perceived demands and available resources.

Anticipatory grief Feelings of loss and guilt while the dying person is still alive.

Anticonformity Opposition to social influence on all occasions, often caused by psychological reactance.

Antidepressant drugs Drugs that alleviate depressive symptoms, presumably because they increase the availability of certain neurotransmitters (especially norepinephrine and serotonin) at synaptic junctions. The three major classes are monoamine oxidase (MAO) inhibitors and tricyclics, and selective serotonin reuptake inhibitors.

Antigen The unique proteins found on the surface of bacteria; these proteins are what enable the immune system to recognize the bacteria as foreign substances.

Antimanic drugs Drugs used to treat bipolar disorder and mania.

Antipsychotic drugs Drugs used to treat psychotic disorders such as schizophrenia.

Antisocial personality A behavioural disorder characterized by a number of deviant acts, including delinquency, truancy, theft, promiscuity, vandalism, fighting, poor work record, impulsiveness, irrationality, reckless behaviour, and aggressiveness.

Antisocial personality disorder A disorder characterized by a failure to conform to standards of decency; repeated lying and stealing; a failure to sustain lasting, loving relationships; low tolerance of boredom; and a complete lack of guilt.

Antlophobia Fear of floods.

Anuptaphobia Fear of staying single.

Anxiety Anxiety is a complex combination of the feeling of fear, apprehension and worry often accompanied by physical sensations such as palpitations, chest pain and/or shortness of breath. It may exist as a primary brain disorder or may be associated with other medical problems including other psychiatric disorders.

Anxiety disorder A mental disorder in which an individual experiences physiological arousal and feelings of tension, tremor, shaking, and general apprehension without obvious reason or provocation.

Anxiety, three types Reality (anxiety about the external world), normal or moral (anxiety about the superego's (originally, the parents') punishing shoulds and oughts), and neurotic (anxiety that a repressed sexual wish might surface). Anxiety is felt only by the ego and might have hereditary components. In his later work Freud referred to anxiety as a danger signal.

Anxiolytics More commonly known as minor tranquilizers, these drugs are given to patients suffering from disabling anxiety. Most types work by increasing the activity of the neurotransmitter GABA and are highly addictive.

Anxious/ambivalent attachment style An expectation about social relationships characterized by a concern that others will not return affection.

Apeirophobia Fear of infinity.

Aphagia Refusal to eat (and in an extreme version, to drink) brought about by a lesion of the lateral hypothalamus.

Aphasia Aphasia is a loss or impairment of the ability to produce or comprehend language, due to brain damage. It is usually a result of damage to the language centres of the brain (like Broca's area). These areas are always located in the left hemisphere and in most people this is where the ability to produce and comprehend language is found. However in a very small number of people language ability is found in the right hemisphere. Damage to these language areas can be caused by a stroke or physical injury. Depending on the area and extent of the damage, someone may be able to speak but not write, or vice versa, or understand more complex sentences than he can produce. The brains of young children with brain damage sometimes restructure themselves to use different areas for speech processing, and regain lost function; adult brains are less "plastic" and lack this ability.

Aphenphosmphobia Fear of being touched. (Haphephobia).

Apiphobia Fear of bees.

Apnea Heavy, disruptive snoring with repetitive pauses in breathing. Results include fatigue, memory loss, impotence, and high blood pressure.

Apotemnophobia Fear of persons with amputations.

Apparent motion A movement illusion in which one or more stationary lights going on and off in succession are perceived as a single moving light; the simplest form of apparent motion is the phi phenomenon.

Apparent movement The perception of movement produced by stimuli that are stationary but flash on and off at appropriate time intervals.

Appeasement display A gesture or pattern of behaviour which signals that an organism is conceding defeat in a conflict.

Appeasement gesture A stereotyped gesture made by a submissive animal in response to a threat gesture by a dominant animal; tends to inhibit an attack.

Appetitive stimulus In instrumental conditioning, a stimulus that the animal will do everything to attain and nothing to prevent.

Applied behaviour analysis The application of behavioural learning principles to understand and change behaviour.

Applied intelligence Practical IQ.

Applied research Research designed to increase the understanding of and solutions to real-world problems by using current social psychological knowledge.

Appraisal The process by which people make judgments about the demands of potentially stressful events and their ability to meet those demands.

Approach approach conflict The conflict presented when two opposite but equally appealing choices are available but can not both be obtained.

Approach avoidance conflict The conflict presented when the best positive choice will result in a negative outcome as well as positive.

Apraxia A serious disturbance in the organization of voluntary action produced by lesions in certain cortical areas, often in the frontal lobes.

Aptitude test A test designed to measure general abilities and to predict future performance.

Aptitude treatment interaction Interaction of individual differences in learning with particular teaching methods.

Arachibutyrophobia Fear of peanut butter sticking to the roof of the mouth.

Arachnephobia or arachnophobia Fear of spiders.

Arbitration Resolution of a conflict by a neutral third party who studies both sides and imposes a settlement.

Archaic heritage Phylogenetic remnants of the species' mental functioning such as inherited dispositions, ideational contents, and memory traces from former generations. Unlike Jung, who coined the term "collective unconscious" (now often referred to as the "objective psyche"), Freud gave these inherited remnants little more than a mention; for him, the wishes, feelings, and aggressions derived from the present life held much more importance. However, he does discuss them as a link between group and individual psychology, a repository of the ancestral memory of the murdered primal father ,and the reason why Oedipal and castration fears are often excessive in comparison to actual family-of-origin dynamics.

Archetypes According to Carl Jung, the stories and images that constitute our collective unconscious.

Architecture for cognition (act*) A model which provides for a complete framework for the 'adaptive control of thought'. The most recent model, act-r (rational act) has a computer implementation that allows experimentation on how people learn to act and think.

Area 17 The primary visual cortex.

Arithmophobia Fear of numbers.

Arousal Arousal is a physiological and psychological state involving the activation of the reticular activating system in the brain stem, the autonomic nervous system and the endocrine system, leading to increased heart rate and blood pressure and a condition of alertness and readiness to respond.

Arousal affect model The proposition that aggression is influenced by both the intensity of arousal and the type of emotion produced by a stimulus.

Arousal theory The theory stating that we are motivated by our innate desire to maintain an optimal level of arousal.

Arousal: cost reward model The proposition that people react to emergency situations by acting in the most cost-effective way to reduce the arousal of shock and alarm.

Arrhenphobia Fear of men.

Arsonphobia Fear of fire.

Artificial intelligence A field that draws on concepts from both cognitive psychology and computer science to develop artificial systems that display some aspects of humanlike intelligence. Examples are computer programs that recognize patterns or solve certain kinds of problems.

Artificial neural network A means of investigating learning in a way that is more 'brain-like' than traditional, symbolic, approaches, and which has many different possible types, each with different learning rules.

Artificial selection A procedure in which animals are deliberately mated to produce offspring that possess particularly desirable characteristics.

Asbergers Also known as Asperger's Disorder or Autistic Psychopathy, is a Pervasive Developmental Disorder (PDD) characterized by severe and sustained impairment in social interaction, development of restricted and repetitive patterns of behaviour, interests, and activities.

Aschematics in experiments In self-concept formation, those people who do not possess a self-schema in a particular domain.

Ascribed characteristics A system for analysing responses to the 'who am I?, which outlines 30 categories of responses under eight broad headings referring to social identity elements. Assimilation-contrast theory one of a variety of social cognition theories which suggests that when other people hold attitudes similar to our own we tend to exaggerate the similarity.

Asexual Not having sexual interests or abilities.

Assertive Style of interpersonal interaction where both the needs of the self and others are considered. (As opposed to passive or aggressive).

Assertive Discipline Giving a clear, firm, unhostile response to student misbehaviour.

Assertiveness Assertiveness is a skill taught by many personal development experts and psychotherapists and the subject of many popular self-help books. It is linked to self-esteem and considered an important communication skill.

Assertiveness training A technique sometimes used by therapists to help patients develop skills in interpersonal relations. In this technique, patients are encouraged to state their needs clearly, negotiate confidently, and persevere when thwarted.

Assessment A measure of the degree to which instructional objectives have been attained.

Assessment center Structured setting in which job applicants are exhaustively tested and judged by multiple evaluators.

Assimilation According to Piaget, the process whereby new cognitive elements are fitted in with old elements or modified to fit more easily; this process works in tandem with accommodation.

Assimilation and accommodation In Piaget's theory, the twin processes by means of which cognitive development proceeds. Assimilation is the process whereby the environment is interpreted in terms of the schemas the child has at the time. Accommodation is the way the child changes his schemas as he continues to interact with the environment.

Assisted suicide Suicide committed with the assistance of a physician by a person terminally ill or in unmanageable pain.

Association A linkage between two psychological processes as a result of past experience in which the two have occurred together. A broad term that subsumes conditioning and association of ideas among others.

Association areas A name sometimes given to regions of the cortex that are not primary projection areas. They tend to be involved in the integration of sensory information or of motor commands.

Association cortex One of several areas in the frontal region of the brain, occupying the temporal, occipital and parietal lobes which receives information from sensory cortical areas as well as other association areas.

Associative links Connections in memory that tie one memory, or one concept, to another.

Associative play Much like parallel play but with increased levels of interaction in the form of sharing, turn-taking, and general interest in what others are doing.

Associative retrieval A type of memory retrieval that seems swift and effortless: The sought-after information simply "pops" into mind.

Asthenophobia Fear of fainting or weakness.

Astraphobia or astrapophobia Fear of thunder and lightning. (Ceraunophobia, Keraunophobia).

Astrophobia Fear of stars or celestial space.

Asymmetriphobia Fear of asymmetrical things.

Atavism (theory of) A classic theory of criminal behaviour that has long been refuted. The theory of atavism suggests that criminals are genetic throwbacks that react deviantly simply because their behaviour matches those of our ancient ancestors.

Ataxiophobia Fear of ataxia. (muscular incoordination).

Ataxophobia Fear of disorder or untidiness.

Atelophobia Fear of imperfection.

Atephobia Fear of ruin or ruins.

Athazagoraphobia Fear of being forgotton or ignored or forgetting.

Atherosclerosis Hardening of arteries because of cholesterol deposits.

Atmosphere hypothesis In syllogistic reasoning, an hypothesis for accounting for errors made when subjects are asked to assess conclusions presented to them, rather than to generate conclusions of their own.

Atomosophobia Fear of atomic explosions.

Attachment A social and emotional bond between infant and caregiver that spans both time and space.

Attachment behaviours A preferential system of approaching and interacting with familiar care givers, also known as proximity-maintaining behaviours, such as crying and clinging, which the infant develops in the first nine months of its life.

Attachment figure The primary care giver recognized by the infant as part of its behavioural system, with whom it maintains proximity by crying or reaching out, and with whom it feels secure.

Attachment style The way a person typically interacts with significant others.

Attachment theory An analytical attempt to explain the various factors that bind people to another in emotional terms, primarily that attachment first established in infancy between the baby and its mother or primary care giver, which is seen as forming the later expectation experienced by the adult in their relationships with others.

Attachment to places In environmental cognition, the development and maintenance of personal and social well-being through identification with particular places.

Attention A collective label for all the processes by which we perceive selectively.

Attention A state of focused awareness on a subset of the available perceptual information.

Attention deficit disorder (ADD) The inability to concentrate for long periods of time.

Attention to the other in the dyad The interpretation of the output from the other.

Attention-deficit/hyperactivity disorder (ADHD) A disorder characterized by difficulties maintaining attention because of a limited ability to concentrate; includes impulsive actions and hyperactive behaviour.

Attitude A fairly stable, evaluative disposition that makes a person think, feel, or behave positively or negatively about some person, group, or social issue.

Attitude accessibility model In the evaluation of social attitudes, an approach which stresses the evaluative component of attitudes as a learned association between an entity and its evaluation, which is stored in memory.

Attitude inoculation Exposing people to weak attacks upon their attitudes so that when stronger attacks come, they will have refutations available.

Attitude scale A multiple-item questionnaire designed to measure a person's attitude toward some object.

Attractiveness Having qualities that appeal to an audience. An appealing communicator (often someone similar to the audience) is most persuasive on matters of subjective preference.

Attribution An interpretive process by which we reach a judgment about the cause(s) of an act or achievement.

Attribution of a disposition In correspondent inference theory, the second of a two-stage process, when observers determine whether a person has a disposition which gives rise to a particular intention.

Attribution of intent In correspondent inference theory, the first of a two-stage process, when observers infer the intention of the person performing a particular behaviour.

Attribution theory Attribution theory is a field of social psychology, which was born out of the theoritical models of Fritz Heider, Harold Kelley, Edward E. Jones, and Lee Ross. Attribution theory is concerned with the ways in which people explain (or attribute) the behaviour of others. It explores how individuals "attribute" causes to events and how this cognitive perception affects their motivation. Think of "explanation" as a synonym and "why" as the question to be answered.

Attributional processes In applied psychology, how individuals assign causal meaning to events.

Attribution-of-arousal theory An approach that combines the James-Lange emphasis on bodily feedback with a cognitive approach to emotion. Various stimuli can trigger a general state of arousal, which is then inter-preted in light of the subject's present situation and shaped into a specific emotional experience.

Atychiphobia Fear of failure.

Atypical antipsychotics Drugs (such as Clozaril, Risperdal, and Zyprexa) that operate by blocking receptors for both dopamine and serotonin; these drugs seem to be effective in treating schizophrenic patients' positive symptoms, such as thought disorders and hallucinations, as well as their negative symptoms, such as apathy and emotional blunting.

Audience design The process of shaping a message depending on the audience for which it is intended.

Audience inhibition effect People are inhibited from helping for fear that other bystanders will evaluate them negatively if they intervene and the situation is not an emergency.

Audition The sense of hearing.

Auditory canal The tube that carries sound from the outer ear to the eardrum.

Auditory cortex The area of the temporal lobes that receives and processes auditory information.

Auditory hair cell The sensory neuron of the auditory system; located on the basilar membrane.

Auditory nerve The nerve that carries impulses from the cochlea to the cochlear nucleus of the brain.

Aulophobia Fear of flutes.

Aurophobia Fear of gold.

Auroraphobia Fear of Northern lights.

Authentic assessment Measurement of important abilities using procedures that simulate the application of these abilities to real-life problems.

Authoritarian [parents] Parenting style focused on excessive rules, rigid belief systems, and the expectation of unquestioned obedience.

Authoritarian personality A cluster of personal attributes (e.g., submission to persons above and harshness to those below) and social attitudes (e.g., prejudice against minority groups) that is sometimes held to constitute a distinct personality.

Authoritative-reciprocal pattern A pattern of child rearing in which parents exercise considerable power but also respond to the child's point of view and reasonable demands. Parents following this pattern set rules of conduct and are fairly demanding but also encourage the child's independence and self-expression.

Autism A category of disability that significantly affects social interaction, verbal and nonverbal communication, and educational performance.

Auto suggestion Self-suggestion.

Autocratic pattern A pattern of child rearing in which the parents control the child strictly, setting stern and usually unexplained rules whose infraction leads to severe, often physical, punishment.

Autodysomophobia Fear of one that has a vile odor.

Autoeroticism Sexual gratification or arousal in the absence of a partner.

Autoerotism Havelock Ellis's term for the baby's habit of pleasing itself sensually with its own body. In normal development it gives way to an erotism directed at love partners.

Autoimmune diseases Diseases such as rheumatoid arthritis, diabetes, lupus, and multiple sclerosis, in which the immune system attacks and destroys some of the body's own tissue.

Autokinetic effect An optical illusion that occurs when someone stares at a stationary point of light in a darkened room where there is no frame of reference. The light appears to move in various directions.

Automatic processes Processes that do not require attention; they can often be performed along with other tasks without interference.

Automatic thoughts Self-defeating judgments people make about themselves.

Automaticity The property of performance of a skill that makes it no longer depend upon effortful, conscious monitoring, such as learning to drive a car or other cognitive skills like reading or playing chess.

Automatism An act performed unconsciously. Defendants have been found innocent due to an automatism defence (i.e., homicide while sleepwalking).

Automatization A process whereby components of a skilled activity become subsumed under a higher-order organization and are run off automatically.

Automatonophobia Fear of ventriloquist's dummies, animatronic creatures, wax statues .

Autonomic arousal The involuntary change in bodily activities that relates to the peripheral nervous system, such as a person's heart rate or sweating, in response to physical or psychological stimuli.

Autonomic nervous system In contrast to the voluntary nervous system, the "involuntary" or autonomic nervous system is responsible for homeostasis, maintaining a relatively constant internal environment by controlling such involuntary functions as digestion, respiration, perspiration, and metabolism, and by modulating blood pressure. Although these functions are generally outside of voluntary control, they are not outside our awareness, and they may be influenced by one's state of mind.

Autonomous morality stage at which a person understands that people make rules and that punishments are not automatic.

Autonomy The ability to function independently without control by others.

Autoreceptor A site located on the synaptic bouton from which monoamine transmitters are released into the synaptic gap of the brain.

Availability One of three types of heuristic, according to which the probability of an event, or of an item having a property, is judged by the ease with which instances can be brought to mind - their availability from memory.

Availability heuristic A rule of thumb often used to make probability estimates, which depends on the frequency with which certain events readily come to mind. This can lead to errors, since, for example, very vivid events will be remembered out of proportion to their actual frequency of occurrence.

Aversion therapy A form of behaviour therapy in which the undesirable response leads to an aversive stimulus (e.g., the patient shocks herself every time she reaches for a cigarette).

Aversive racism Attitudes toward members of a racial group that incorporate both egalitarian social values and negative emotions, causing one to avoid interaction with members of the group.

Aversive stimulus In instrumental conditioning, a stimulus such as an electric shock, which the animal does everything to avoid and nothing to attain.

Avoidance avoidance conflict The conflict where both possible choices have an equal negative outcome.

Avoidance behaviour Behaviour that results in the removal of an ongoing event, or prevents a future event from occurring.

Avoidance learning Instrumental learning in which the response averts an aversive stimulus before it occurs.

Avoidance response An operant response acquired through negative reinforcement that prevents an aversive stimulus from occurring.

Avoidant attachment style An expectation about social relationships characterized by a lack of trust and a suppression of attachment needs.

Avoidant personality disorder A disorder involving social withdrawal, low self-esteem, and extreme sensitivity to being evaluated negatively.

Axon A long, thin part of a neuron attached to the soma; divides into a few or many branches, ending in terminal buttons.

Axon terminals The knoblike swellings on the ends of an axon. The terminals contain the synaptic vesicles that are filled with neurotransmitters.

B

B lymphocytes Cells that develop in bone marrow and release immunoglobulins to defend the body against antigens.

Babble The first stage in the infant's progress towards speech production, highlighted by the playful production of universally recognizable single-consonant and vowel-like sounds and repetitive syllable strings.

Bacillophobia Fear of microbes.

Background Firing rate most neurones are not 'silent', but respond spontaneously with a background, or resting, rate of action potentials, even when not directly stimulated.

Backward masking The ability of a stimulus to interfere with the perception of a stimulus presented just before it.

Backward pairing A classical conditioning procedure in which the conditioned stimulus (CS) follows the unconditioned stimulus (US).

Backward planning Planning instruction by first setting long-range goals, then setting unit objectives, and finally planning daily lessons.

Bacteria Tiny creatures making up a division (Bacteria) of microorganisms that are typically one-celled, have no chlorophyll, multiply by simple division, and can be seen only with a microscope.

They occur in three main forms, spherical (cocci), rod-shaped (bacilli), and spiral (spirilla); some bacteria cause diseases such as pneumonia and anthrax, and others are necessary for fermentation, nitrogen fixation, and so on.

Bacteriophobia Fear of bacteria.

Balance theory A theory that people desire cognitive consistency or balance in their thoughts, feelings, and social relationships.

Balint's syndrome A syndrome caused by bilateral damage to the parieto-occipital region of the brain; includes difficulty in perceiving the location of objects and reaching for them under visual guidance.

Ballistics Ballistic experts focus on the functioning of firearms. Via microscopic analysis they can match up bullets with a particular weapon. They also provide key information about the projectiles path.

Ballistophobia Fear of missiles or bullets.

Ballottement A type of pelvic examination in which a physician or nurse feels for a foetus in the uterus.

Barbiturate A drug that causes sedation; one of several derivatives of barbituric acid.

Bargaining One of Elisabeth Kubler-Ross's stages of dying in which the dying tries to bargain with God or another religious figure, promising to change, make amends, or atone for his or her wrongdoings in order to avoid death.

Barnum effect Describes the fact that a description of one's personality will often be uncritically accepted as valid if it is stated in sufficiently general terms.

Barophobia Fear of gravity.

Basal ganglia A collective name for a set of structures located around the thalamus, also known as the putamen and globus pallidus, that are part of the motor control mechanism of the brain.

Basal metabolic rate The rate at which energy is used when a person is at complete rest.

Base rates Those rates of occurrence of a phenomenon which need to be taken into account in order to determine the diagnosticity of evidence.

Base-rate fallacy The failure to consider the likelihood that a person, place, or thing is a member of a particular category.

Basic Intelligence academic IQ.

Basic emotions According to some theorists, a small set of elemental, built-in emotions revealed by distinctinve patterns of physiological reaction and facial expression.

Basic level The level of categorization that can be retrieved from memory most quickly and used most efficiently.

Basic orientations Horney's sets of personality characteristics that correspond to the strategies of moving toward others, moving against others, and moving away from others.

Basic research Research designed to increase knowledge about social behaviour.

Basic rest-activity cycle (BRAC) A 90-minute cycle (in humans) of waxing and waning alertness controlled by a biological clock in the pons; during sleep, it controls cycles of REM sleep and slow-wave sleep.

Basic-level concept A concept that makes important distinctions between different categories.

Basilar membrane A membrane in the cochlea that, when set into motion, stimulates hair cells that produce the neural effects of auditory stimulation.

Basking in reflected glory (BIRGing) Actively identifying with and embracing the success and positive evaluations of others as is they were one's own.

Basophobia or basiphobia Inability to stand. Fear of walking or falling.

Bathmophobia Fear of stairs or steep slopes.

Bathophobia Fear of depth.

Batophobia Fear of heights or being close to high buildings.

Batrachophobia Fear of amphibians, such as frogs, newts, salamanders, etc.

Before and after Designs with a pretest and posttest but no separate control group—the subject is its own control based on previous behaviour.

Behaviour The actions by which an organism adjusts to its environment.

Behaviour analysis A branch of psychology that studies the effect of the environment on behaviour; primarily, the effects of the consequences of behaviours on the behaviours themselves.

Behaviour content matrix A chart that classifies lesson objectives according to cognitive level.

Behaviour genetics The study of behaviour and personality differences among people.

Behaviour modification A common psychological treatment used in both clinical and forensic environments. The changing of human behaviour by the application of conditioning and/or other learning techniques.

Behaviour therapy A general approach to psychological treatment which 1. holds that the disorders to which it addresses itself are produced by maladaptive learning and must be remedied by reeducation,

2. Proposes techniques for this reeducation based on principles of learning and conditioning,

3. Focuses on the maladaptive behaviours as such rather than on hypothetical unconscious processes of which they may be expressions.

Behavioural approach system (BAS) Brain system that underlays the tendency to seek out pleasant stimuli.

Behavioural confirmation The process by which people behave in ways that elicit from others specific expected reactions and then use those reactions to confirm their beliefs.

Behavioural contrast A pattern of responding in which an organism seems to evaluate a reward relative to other rewards that are available or that have been available recently. For example, an animal might respond only weakly to a reward of two pellets if it recently received a reward of five pellets for some other response.

Behavioural data Observational reports about the behaviour of organisms and the conditions under which the behaviour occurs or changes.

Behavioural genetics A subfield of psychology that examines the role of genetic factors in behaviour.

Behavioural inhibition system (BIS) Brain system that underlays the tendency to avoid unpleasant stimuli.

Behavioural learning theory Explanation of learning that emphasizes observable changes in behaviour.

Behavioural measures Overt actions and reactions that are observed and recorded, exclusive of self-reported behaviour.

Behavioural medicine An interdisciplinary field that integrates and applies behavioural and medical knowledge about health and disease.

Behavioural pharmacology The study of how drugs influence behaviour; combines the principles of operant conditioning with the principles of drug action.

Behavioural rehearsal Procedures used to establish and strengthen basic skills; as used in social-skills training programs, requires the client to rehearse a desirable behaviour sequence mentally.

Behavioural-cognitive approach to personality An approach that defines personality differences by the way in which different people act and think about their actions. It tends to emphasize situational determinants and prior learning in trying to explain how such differences come about.

Behaviourism A framework for understanding human behaviour through observable, measurable data. This view emphasizes objective stimulus and response over more subjective analysis of internal states. Key figures in behaviourism include American psychologists John B. Watson (1878-1958) and B.F. Skinner (1904-1990).

Behaviourist perspective The psychological perspective primarily concerned with observable behaviour that can be objectively recorded and with the relationships of observable behaviour to environmental stimuli.

Behaviour setting In environmental psychology, those aspects of a sociophysical environment with which the individual interacts, such as other people and inanimate objects.

Behavioural intention In the explanation of social attitudes, the tendency to act towards the entity in a particular way.

Belief An estimate of the probability that something is true.

Belief in a just world The belief that individuals get what they deserve in life, an orientation that leads people to disparage victims.

Belief perserverance Persistence of one's initial conceptions, as when the basis for one's belief is discredited but an explanation of why the belief might be true survives.

Belief-bias effect A situation that occurs when a person's prior knowledge, attitudes, or values distort the reasoning process by influencing the person to accept invalid arguments.

Bell shaped curve Also referred to as a normal distribution or normal curve, a bell-shaped curve is a perfect mesokurtic curve where the mean, median, and mode are equal.

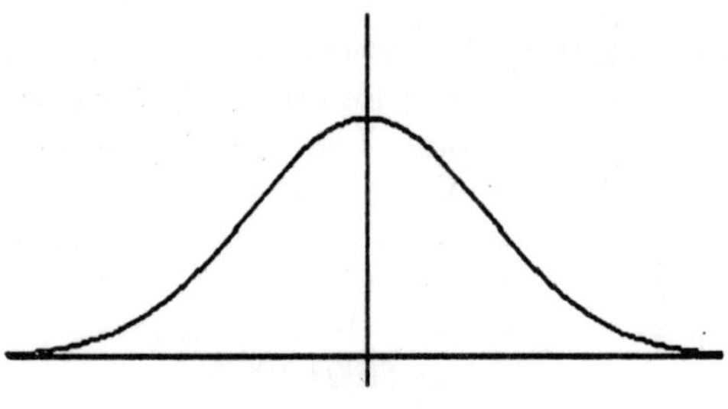

Fig. Bell shaped curve

Belonephobia Fear of pins and needles. (Aichmophobia).

Belongingness in learning The fact that the ease with which associations are formed depends upon the items to be associated. This holds for classical conditioning in which some CS-US combinations are more effective than others (e.g., learned taste aversions) and for instrumental conditioning in which some response-reinforcer combinations work more easily than others (e.g., specific defence reactions in avoidance conditioning of species).

Benzodiazepines A class of antianxiety drugs. They are also called tranquilizers.

Bereavement To be left in a sad or lonely state, as by loss or death.

Beta activity The irregular, high-frequency activity of the electroencephalogram, usually indicating a state of alertness or arousal.

Beta rhythm A rhythmic pattern in the electrical activity of the brain, often observed when one is engaged in active thought.

Beta waves The type of brain waves present when a person is awake and alert.

Between-class ability grouping The practice of grouping students by ability level in separate classes within-class ability.

Between-family differences In research on the genetics of behaviour, a term often used to refer to the role of environment. It describes environmental differences that apply to entire families, such as differences in socioeconomic status, religion, or child-rearing attitudes. For most personality attributes, these seem to be less important than within-family differences.

Between-group heritability The extent to which variation between groups (as in the difference between the mean iqs of U.S. whites and blacks) is attributable to genetic factors.

Between-subjects design A research design in which different groups of participants are randomly assigned to experimental conditions or to control conditions.

Bias In the consideration of arguments, an inability to allow all possible arguments about an issue, thus considering mainly or only the arguments on one side of a particular case.

Bibliophobia Fear of books.

Bibliotherapy Dealing with death by reading books about dying.

Bidirectional activation models Models of pattern recognition in which elements are activated as well as inhibited from both lower levels (bottom-up processing) and higher levels (top-down processing).

Big Five Five basic personality traits from which other traits are derived. They include neuroticism, extraversion, openness to experience, agreeableness, and conscientiousness.

Bilingual Using or capable of using two languages, especially with equal or nearly equal facility.

Bilingual education Instructional program for students who speak little or no English in which some instruction is provided in the native language.

Bimodal A frequency curve in which there are two modes, resulting in a "dip" in the middle of the graph. The mean and median fall between the two modes.

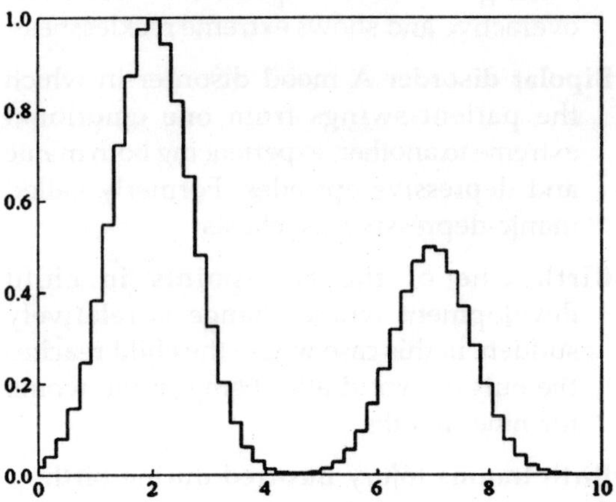

Fig. Bimodal

Binaural comparison Comparison of the responses in the two ears, which is particularly important in location the direction of sound sources.

Bind the process by which the secondary process manages or traps free libido. Thinking instead of fornicating, for instance. (Obviously, binding has disadvantages as well as advantages....).

Binding problem The problem confronted by the brain of recombining the elements of a stimulus, once these elements have been separately analyzed by different neural systems.

Binet-Simon scale In intelligence testing, a system of 30 tests in a scale of increasing difficulty as a way of measuring the child's scholastic performance in comparison with expected levels of performance of other children of the same age.

Binocular cues Visual cues (convergence and retinal disparity) that require both eyes to perceive distance (as opposed to monocular cues).

Binocular disparity An important cue for depth perception. Each eye obtains a different view of an object, the disparity becoming less pronounced the farther the object is from the observer.

Binocular stereopsis Literally 'two-eyed sold vision', this is the process of using binocular disparity to recover information about the relative distance of objects.

Binocularly-driven cell A cell with a receptive field in each of the eyes, which will thus respond to appropriate stimulation of either eye.

Bio feedback System to enable people to become consciously aware of and control vital physiological functions involving measurement of the function, amplification,

Monitoring and continuous presentation through a sensory stimulus to the person.

Biochemistry Biochemistry is the chemistry of life. Biochemists study the molecules and chemical reactions catalyzed by enzymes that take place in all living organisms. See the article on molecular biology for a diagram and description of the interrelationship between biochemistry, molecular biology, and genetics.

Biological biasing The idea that people are genetically "primed" for a disorder, and therefore more likely to get it than others in the general population.

Biological constraints on learning Any limitations on an organism's capacity to learn that are caused by the inherited sensory, response, or cognitive capabilities of members of a given species.

Biological evolution Changes in the genetic or physical characteristics of a population or group of organisms over time.

Biological motion The stimulus motion produced by animate things which is most often used to refer to particular moving stimuli, where the movement of a person in represented only by a few lights attached to the major joints.

Biological perspective The approach to identifying causes of behaviour that focuses on the functioning of the genes, the brain, the nervous system, and the endocrine system.

Biological preparedness Evolved predisposition to learn a behavioural contingeny.

Biological rhythms Periodic physiological changes.

Biological senescing The process of growing older physically.

Biomedical approach A psychological approach to abnormality which emphasizes both qualitative and quan-titative differences in order to ascertain what is underlying a particular symptom or deviation of a diagnostic sign.

Biomedical model An approach to mental disorders that emphasizes somatogenic causes.

Biomedical therapies Treatments for psychological disorders that alter brain functioning with chemical or physical interventions such as drug therapy, surgery, or electroconvulsive therapy.

Biopsychosocial Perspective studies human development by examining the interaction of biological, psychological, and social factors.

Biopsychosocial model A model of health and illness that suggests that links among the nervous system, the immune system, behavioural styles, cognitive processing, and environmental factors can put people at risk for illness.

Bipedalism The ability to move about the environment on two feet.

Bipolar cell A neuron in the retina that receives information from photoreceptors and passes it on to the ganglion cells, from which axons proceed through the optic nerves to the brain.

Bipolar depression Also known as manic depression, a severe form of depression where the sufferer alternates between periods of depression and mania, and during the manic phase is irritable, overactive and shows extreme recklessness.

Bipolar disorder A mood disorder in which the patient swings from one emotional extreme to another, experiencing both manic and depressive episodes. Formerly called manic-depressive psychosis.

Birth One of the two points in child development where change is relatively sudden, in this case where the child reaches the outside world after being in the womb for nine months.

Birth trauma Injury incurred during birth.

Bisexuality A sexual orientation in which a person has erotic and romantic feelings for both their own and the opposite sex.

Blank lineup In order to assess the accuracy of a witness police officers may present a police lineup that does not contain the suspect.

Blastocyst An embryo at the stage of development in which it consists of usually one layer of cells around a central cavity, forming a hollow sphere.

Blended families Stepfamilies, in which new family units are made up of children from previous marriages.

Blennophobia Fear of slime.

Blind spot The region of the eye that contains no visual receptors and therefore cannot produce visual sensations.

Blind study As a way to avoid the placebo effect in research, this type of study is designed without the subject's knowledge of the anticipated results and sometimes even the nature of the study. The subjects are said to be 'blind' to the expected results.

Blindsight The ability of a person with a lesion in the visual cortex to reach toward or guess at the orientation of objects projected on the part of the visual field that corresponds to this lesion, even though they report that they can see absolutely nothing in that part of their visual field.

Blocking A phenomenon in which an organism does not learn a new stimulus that signals an unconditioned stimulus, because the new stimulus is presented simultaneously with a stimulus that is already effective as a signal.

Blocking effect An effect produced when two conditioned stimuli, *A* and *B*, are both presented together with the unconditioned stimulus (US). If stimulus *A* has previously been associated with the unconditioned stimulus while *B* has not, the formation of an association between stimulus *B* and the US will be impaired (that is, blocked).

Blood-brain barrier Specialized membranes that surround the blood vessels within the brain and that filter toxins and other harmful chemicals, ensuring brain cells' a relatively pure blood supply.

Body esteem A person's attitudes toward his or her body.

Body-image In self-descriptions, a person's perceptions of their own body, which may involve either neutral descriptions relating to height, weight, etc., or evaluative judgements of appearance.

Bogus pipeline A procedure that fools people into disclosing their attitudes. Participants are first convinced that a machine can use their psychological responses to measure their private attitudes. Then they are asked to predict the machine's reading, thus revealing their attitudes.

Bogyphobia Fear of bogeys or the bogeyman.

Bolshephobia Fear of Bolsheviks.

Borderline personality disorder A disorder characterized by impulsive behaviour and unstable relationships, emotions, and self-image.

Botanophobia Fear of plants.

Bottom-up processing A type of information processing which is thought to progress from a simple description of the stimulus (e.g. The image) to a more complex description of the external world (i.e. A perceptual model).

Brain The main organ in the nervous system.

Brain lesion Damage to a particular region of the brain.

Brain stem The "stem" of the brain, including the medulla, pons, and midbrain.

Brain waves Tracings that show the electrical activity of the brain.

Brainstorming A technique that attempts to increase the production of creative ideas by encouraging group members to speak freely without criticizing their own or others' contributions.

Breech presentation The delivery of a foetus presenting itself with its buttocks or feet first.

Brightness A perceptual dimension of colour, most closely related to the intensity or degree of radiant energy emitted by a visual stimulus.

Brightness constancy The tendency to perceive objects as having constant brightness even when they are observed under varying levels of illumination.

Brightness contrast The perceiver's tendency to exaggerate the physical difference in the light intensities of two adjacent regions. As a result, a gray patch looks brighter on a black background, darker on a white background.

Brightness ratio The ratio between the light reflected by a region and the light reflected by the area that surrounds it. According to one theory, perceived brightness is determined by this ratio.

Broca's aphasia A condition, described by the French surgeon, Paul Broca, where people have slow and stilted speech as a result of a dysfunction in the precentral gyrus responsible for organizing speech output.

Broca's area A part of the brain, in the left frontal lobe, that is involved in speech production.

Bromidrosiphobia or Bromidrophobia Fear of body smells.

Brontophobia Fear of thunder and lightning.

Buddhism Religion founded by Gautam Buddha in 6th century B.C. which is very much in line with modern scientific outlook and method.

Bufonophobia Fear of toads.

Bulimia nervosa A disorder involving binge eating followed by compensatory behaviours such as vomiting, fasting, excessive exercise, or use of laxatives, diuretics, and other medications to control body weight.

Bullying Persistent aggressive behaviour directed towards a particular victim, particularly children at school, who cannot retaliate effectively.

Bullying persistent Aggressive behaviour directed towards a particular victim, particularly children at school, who cannot retaliate effectively.

Burnout Changes in thoughts, emotions, and behaviour as a result of extended job stress and unrewarded repetition of duties. Burnout is seen as extreme dissatisfaction, pessimism, lowered job satisfaction, and a desire to quit.

Bystander effect The phenomenon that underlies many examples of failing to help strangers in distress: The larger the group a person is in (or thinks he is in), the less likely he is to come to a stranger's assistance. One reason is diffusion of responsibility (no one thinks it is *his* responsibility to act).

Bystander intervention model A theory that whether bystanders intervene in an emergency is a function of a 5-step decision making process.

C

Cacophobia Fear of ugliness.

Cainophobia or cainotophobia Fear of newness, novelty.

California Psychological Inventory (CPI) A commonly used personality test, aimed especially at high-school and college students, that tests for traits such as dominance, sociability, responsibility, and so on.

Caligynephobia Fear of beautiful women.

Calling order The order in which students are called on by the teacher to answer questions asked during the course of a lesson.

Cancerophobia or carcinophobia Fear of cancer.

Cannon-Bard theory of emotion A theory stating that an "emotional stimulus produces two co-occurring reactions—arousal "and experience of emotion—that do not cause each other.

Cannula A tiny tube used to inject or withdraw small quantities of brain chemicals.

Canonical correlation A correlational technique used when there are two or more X and two or more Y. (Example: The correlation between (age and sex) and (income and life satisfaction).

Cardiophobia Fear of the heart.

Carnophobia Fear of meat.

Case grammar A form of semantic grammar which represents the semantic roles of phrases using semantic cases such as agents, objects and instruments.

Case study A detailed study of one particular example of a phenomenon which involves in-depth analysis of a particular individual's behaviour.

Case study research Research in which an investigator studies an individual who has a rare or unusual condition or who has responded favourably to a new treatment.

Castration The surgical removal of the testes or ovaries. In the past was used as a treatment for male sex offenders. The ethicity and validity of this approach has come into question.

Castration anxiety What boys fear from their fathers, a fear reinforced by the boy's eventual knowledge that women have no penis and are therefore "castrated." Castration anxiety forces the boy to repress his sexual desire for his mother and identify with his father, a formal rival for mother's love, and grow up to smoke a lot of cigars by way of defensive compensation.

Castration fear (boys) and penis envy (girl) together make up the "castration complex." Because women feel less intense castration anxiety (being "castrated" already), they develop less of a superego.

Catagelophobia Fear of being ridiculed.

Catalepsy A trancelike state with loss of voluntary motion and failure to react to stimuli.

Catapedaphobia Fear of jumping from high and low places.

Cataplexy A neurological disorder in which the person collapses, becoming temporarily paralyzed but not unconscious; usually triggered by anger or excitement; apparently related to the paralysis that normally accompanies REM sleep.

Catatonic Total withdrawal characterized by physical and mental rigidity and inactivity.

Catatonic schizophrenia A subcategory of schizophrenia. Its main symptoms are peculiar motor patterns, such as periods in which the patient is immobile and maintains strange positions for hours on end.

Catatonic type A subtype of schizophrenia characterized by unnatural movement patterns such as rigid, unmoving posture or continual, purposeless movements, or by unnatural speech patterns such as absence of speech or parroting of other people's speech.

Catch trials Trials in a signal detection experiment in which no signal is presented. These trials ensure that the observer is taking the task seriously and truly trying to determine whether a signal is present or not.

Catecholamines A family of neurotra-nsmitters that have an activating function, including epinephrine, norepine-phrine, and dopamine.

Categorical scale A scale that divides responses into categories that are not numerically related.

Category An aspect of a variable—e.g. "male" and "female" if the variable is gender.

Catharsis The emotional release associated with the expression of unconscious conflicts.

Catharsis hypothesis The theory that states that if angry persons can express aggression in a safe manner, they will be less likely to engage in harmful behaviour. Has possible implications with offender treatment.

Cathartic method The name Freud and Joseph Breuer gave to their method of allowing patients to get relief by talking out their previously repressed emotions. Freud quickly realized that this relief was only temporary and did not produce lasting personality changes. Breuer ran away when a patient fell in love with him. Freud lit a cigar.

Cathexis The investment of libido in objects. An example would be Freud's enormous cathexis of interest around sexuality. Cathexes correspond to ideas, whereas affects are discharge products.

Note: in German the noun Freud uses is.

Besetzung and the verb is besetzen: occupation, to occupy. Given Freud's fondness for mechanico-electric metaphors, a more accurate word than cathexis might be charge.

Cathisophobia Fear of sitting.

Catoptrophobia Fear of mirrors.

Causal event An event that causes another event to occur.

Causal relationship A relationship between two variables in which one variable causes the other. This relationship is hard to discover with absolute certainty.

CEEB score A standard score that sets the mean to five-hundred and standard deviation to one-hundred. Used on the Scholastic Aptitude Test (SAT).

Celibate Abstaining from sexual intercourse.

Cell The cell is the structural and functional unit of all living organisms, sometimes called the "building blocks of life." Some organisms, such as bacteria, are unicellular, consisting of a single cell. Other organisms, such as humans, are multicellular, (humans have an estimated 100,000 billion = 1014 cells). The cell theory, first developed in the 19th century, states that all organisms are composed of one or more cells; all cells come from preexisting cells; all vital functions of an organism occur within cells and that cells contain the hereditary information necessary for regulating cell functions and for transmitting information to the next generation of cells.

Cell body The main part of a neuron where the information is processed.

Cenophobia or centophobia Fear of new things or ideas.

Central fissure The visible fissure in the brain that separates the frontal and parietal lobes.

Central limit theorem States: The means of samples drawn from a population will be distributed normally.

Central nervous system (CNS) One of the two major divisions of the body's nervous system, which is comprised of the brain and the spinal cord.

Central pattern generators) Circuits in the nervous system that orchestrate lower level reflexes and other neural activities into larger, organized acts. CPGS instigate certain crucial basic actions, such as, chewing, breathing, locomotion, etc.

Central route to persuasion The path of persuasion when the issue being discussed matters to the listener and the listener is not distracted. Persuasiveness will be increased by convincing arguments; other considerations (whether the person offering the arguments is attractive, the number of arguments offered) are less important.

Central tendency The tendency of scores in a frequency distribution to cluster around a central value.

Central trait A trait that is associated with many other attributes of the person who is being judged. Warmth and coldness are central because they are important in determining overall impressions.

Centration A thought pattern common during the beginning of the preoperational stage of cognitive development; characterized by the child's inability to take more than one perceptual factor into account at the same time.

Centre subregion One of two concentric, circular regions which make up the receptive fields of retinal ganglion cells and which are of opposite sign (excitatory or inhibitory) and exhibit lateral inhibition.

Cephalocaudal order The order in which foetal development occurs, beginning with the head and ending with the lower body and extremities.

Ceraunophobia or keraunophobia Fear of thunder and lightning.(Astraphobia, Astrapophobia).

Cerebellum The cerebellum (literally "little brain") is a brain region important for the integration of sensory perception with motor output. The numerous loops both within and through the cerebellum with the motor cortex and spinocerebellar tracts indicate the cerebellum is as an integrative region, modulating function. Lesions of the cerebellum do not cause paralysis but rather cause feedback deficits manifesting as disorders in fine movement, equilibrium, posture, and motor learning.

Cerebral cortex The outer layer of the cerebral hemispheres of the brain, approximately 3 mm thick.

Cerebral hemisphere The cerebral hemisphere forms one half of a brain. Humans (and many other types of animals) have a brain divided into two hemispheres. Each hemisphere has corresponding structures, and structural locations; however, they are not mirror images. Instead, the hemispheres usually are asymmetrical giving specialized function to each hemisphere. Each hemisphere also has an outer layer of grey matter called the cerebral cortex.

Cerebral palsy Disorder in ability to control movements caused by damage to the motor area of the brain.

Cerebro-spinal fluid A plasma filtrate that is secreted from the capillaries of the ventricles of the brain and other blood vessels and circulates around these and other cavities.

Cerebrum The two major hemispheres of the brain, which dominate its external appearance and control all voluntary motor activity and higher mental functions.

Cesarean section (C-section) A surgical operation for delivering a baby by cutting through the mother's abdominal and uterine walls.

Chaetophobia Fear of hair.

Channel of communication The way the message is delivered-whether face to face, in writing, on film, or in some other way.

Cheimaphobia or cheimatophobia Fear of cold.(Frigophobia, Psychophobia).

Chemical imbalance theory The chemical imbalance theory posits a belief there is a simple chemical or physical malfunction of the brain, causing mental disorders, which can be fixed using psychotropic drugs, electroconvulsive therapy, or psychosurgery. The term 'chemical imbalance', used extensively as a tagline by drug companies in the United States after deregulation of pharmaceutical advertising, is not a medical or scientific term, and in this respect, is not the basis for any scientific theory. The term 'chemical imbalance theory' is most frequently used by critics of the drug industry.

Chemophobia Fear of chemicals or working with chemicals.

Chemosense One of the two sense modalities (gustation and olfaction) that detect the presence of particular molecules present in the environment.

Cherophobia Fear of gaiety.

Chi square one way A test that determines whether there is a meaningful relationship between two nominal variables.

Chi square two way a test that determines whether there is a meaningful relationship between two nominal variables that each have more than one level.

Child development The maturation of children.

Child maltreatment/abuse Actions either physical or psychological that harm children. Can be either a voluntary or involuntary action.

Child molestation The sexual abuse of a child, which occurs when a teenager or adult entices or forces a child to participate in sexual activity.

Child physical abuse The intentional infliction of pain, injury, and harm onto a child.

Child-directed speech (CDS) A term, formerly called motherese, to describe the short, fluent, grammatically correct utterances, with enhanced intonation, by which mothers, other adults and older children help toddlers to learn language.

Childhood amnesia The failure to remember the events of our very early childhood. This is sometimes ascribed to massive change in retrieval cues, sometimes to different ways of encoding memories in early childhood.

Chionophobia Fear of snow.

Chiraptophobia Fear of being touched.

Chirophobia Fear of hands.

Chloasma A skin discolouration on the face and chest, resulting from pregnancy, disease, malnutrition, and so on.

Choice reaction time A measure of the speed of mental processing in which the subject has to choose between one of several responses depending upon which stimulus is presented.

Cholecystokinin (CCK) A hormone released by the duodenum that appears to send a "stop eating" message to the brain.

Cholerophobia Fear of anger or the fear of cholera.

Choral response A response to a question made by an entire class in unison.

Chorion The outermost of the two membranes that completely envelop a foetus.

Chorionic villi Small fingerlike projections in the placenta through which foetal blood circulates.

Chorionic villus sampling A test for detecting genetic abnormalities, determining sex, and so on in a foetus.

Tissue samples of chorionic villus are removed from the uterus.

Chorophobia Fear of dancing.

Chromatic colours Colours that have a discernible hue. These are in contrast to the achromatic colours, which include black, the various shades of grey, and white.

Chromatopsia The inability to discriminate among different hues; caused by damage to the visual association cortex.

Chrometophobia or chrematophobia Fear of money.

Chromophobia or chromatophobia Fear of colours.

Chromosomal aberration The rearrangement of genes within cells or a change in the total number of chromosomes.

Chromosomes Structures in the nucleus of each cell that contain the genes, the units of hereditary transmission. A human cell has forty-six chromosomes, arranged in twenty-three pairs. One of these pairs consists of the sex chromosomes. In males, one member of the pair is an X-chromosome, the other a Y-chromosome. In females, both members are X-chromosomes.

Chronic stress A continuous state of arousal in which an individual perceives demands as greater than the inner and outer resources available for dealing with them.

Chronological age The age of an individual in years.

Chronomentrophobia Fear of clocks.

Chronophobia Fear of time.

Chunk A unit of information used in memory.

Chunking A process by which information is simplified by rules that make it easier to remember. For example, the string of letters TWAABCFBI is easier to remember if a person learns the rule that organizes it into smaller "chunks": TWA, ABC, and FBI.

Cibophobia Fear of food.(Sitophobia, Sitiophobia).

Cilium A hairlike appendage of a cell; involved in movement or in transducing sensory information. Cilia are found on the receptors in the auditory and vestibular system.

Cingulotomy Surgical destruction of the cingulum bundle, which connects the prefrontal cortex with the limbic system;

helps to reduce intense anxiety and the symptoms of obsessive-compulsive disorder.

Circadian rhythm A rhythm that spans about a twenty-four-hour day, such as that of the sleep-waking cycle. Circadian rhythms in humans originate from a clock circuit in the hypothalamus that is set by information from the optic nerve about whether it is day or night.

Circumplex Circular arrangement of variables in two-dimensional space, including all possible blends of the two dimensions.

Cisterna A structure which contains the synaptic vesicles and from which bits of vesicle membrane are pinched off.

Cistron A piece of dna, often also called a gene.

Cladistic taxonomy A classification scheme applied to primates, which uses the possession of shared, derived features as a way of defining groups.

Class inclusion A skill learned during the concrete operational stage of cognitive development in which individuals can think simultaneously about a whole class of objects as well as relationships among its subordinate classes.

Classical antipsychotics Drugs (such as Thorazine and Haldol) that operate by blocking receptors for dopamine; these drugs seem to be effective in treating many schizophrenic patients' positive symptoms, such as thought disorders and hallucinations. Also called major tranquillizers and neuroleptics.

Classical conditioning Classical conditioning, also called "Pavlovian conditioning" or "respondent conditio-ning", is a type of learning found in animals, caused by the association (or pairing) of two stimuli. The simplest form of classical conditioning is reminiscent of what Aristotle would have called the law of contiguity. Essentially, Aristotle said, "When two things commonly occur together, the appearance of one will bring the other to mind.".

Classical psychoanalysis The method developed by Sigmund Freud which assumes that a patient's ills stem from unconscious defences against unacceptable urges that date back to early childhood.

Classification The ability to group according to features.

Classroom management Methods used to organize classroom activities, instruction, physical structure, and other features to make effective use of time, to create a happy and productive learning environment, and to minimize behaviour problems and other disruptions.

Claustrophobia Fear of confined spaces.

Cleithrophobia or cleisiophobia Fear of being locked in an enclosed place.

Cleptophobia Fear of stealing.

Client The term used by clinicians who think of psychological disorders as problems in living, and not as mental illnesses, to describe those being treated.

Client centred therapy Also known as person centred or Rogerian therapy, a form of therapy which brings the client's personal needs into action, viewing the client as intrinsically good and offering a sense of unconditional acceptance and regard for them.

Climacophobia Fear of stairs, climbing, or of falling downstairs.

Clinical ecology A field of psychology that relates disorders such as anxiety and depression to environmental irritants and sources of trauma.

Clinical psychologist An individual who has earned a doctorate in psychology and whose training is in the assessment and treatment of psychological problems.

Clinical psychology The branch of psychology devoted to the investigation and treatment of abnormal behaviour and mental disorders.

Clinical social worker A mental health professional whose specialized training prepares him or her to consider the social context of people's problems.

Clinophobia Fear of going to bed.

Clithrophobia or Cleithrophobia Fear of being enclosed.

Closure A factor in visual grouping. The perceptual tendency to fill in gaps in a figure so that it looks complete.

Cluster sampling The researcher subdivides an expansive area into smaller units and then a subset of the identified clusters is randomly selected. Subjects within clusters are related in some meaningful way.

Cnidophobia Fear of stings.

Co sleeping children Sleeping in the same bed as their parents.

Co-actor One of two individuals performing the same task within the same space. Cochlea the main organ of the inner ear, responsible for the initial analysis of sound waveforms.

Cochlea A coiled structure in the inner ear that contains the basilar membrane whose deformation by sound-produced pressure stimulates the auditory receptors.

Cocktail party phenomenon The ability to attend to a single conversation in a crowded room; an example of stimulus segmentation.

Cocktail-party effect The effect one experiences in settings such as noisy parties, where one tunes in to the voice of the person one is talking to and filters out the other voices as background noise. This phenomenon is often taken as the model for studying selective attention based on listening to speech.

Coding The translation of stimulus information into various dimensions of sensation (e.g., intensity and quality) that are actually experienced.

Codon A triplet of bases along dna which build amino acids.

Coefficient of determination The statistic or number determined by squaring the correlation coefficient. Represents the amount of variance accounted for by that correlation.

Coefficient of relatedness In sociobiology, the average proportion of genes that are shared between family members.

Coercion Coercion is the practice of compelling a person to act by employing threat of harm (usually physical force, sometimes other forms of harm). Often, it involves the use of actual force in order to make the threat credible, but it is the threat of (further) force which brings about cooperation of the person being coerced.

Coercive power Power derived through the ability to punish.

Cognition Processes of knowing, including attending, remembering, and reasoning; also the content of the processes, such as concepts and memories.

Cognitive appraisal With respect to emotions, the process through which physiological arousal is interpreted with respect to circumstances in the particular setting in which it is being experienced; also, the recognition and evaluation of a stressor to assess the demand, the size of the threat, the resources available for dealing with it, and appropriate coping strategies.

Cognitive appraisal theory of emotion A theory stating that the experience of emotion is the joint effect of physiological arousal and cognitive appraisal, which serves to determine how an ambiguous inner state of arousal will be labeled.

Cognitive apprenticeship The process by which a learner gradually acquires expertise in interaction with an expert, either an adult or an older or more advanced peer.

Cognitive behaviour modification A therapeutic approach that combines the cognitive emphasis on the role of thoughts and attitudes influencing motivations and response with the behavioural emphasis on changing performance through modification of reinforcement contingencies.

Cognitive behavioural therapy Treatment involving the combination of behaviourism (based on the theories of learning) and cognitive therapy (based on the theory that our cognitions or thoughts control a large portion of our behaviours).

Cognitive components The mental processes needed to solve complex problems. Some theories of intelligence propose that individuals differ in their skill in using these component mental processes.

Cognitive consistency A state in which one's beliefs and preferences are consistent with each other.

Cognitive control The ability to create subjective realities for oneself, or as directed by a leader; the power of an individual or group to give different meanings to situations.

Cognitive development Gradual, orderly changes by which mental processes become more complex and sophisticated.

Cognitive dissonance Tension that arises when one is simultaneously aware of two inconsistent cognitions. For example, dissonance may occur when we realize that we have, with little justification, acted contrary to our attitudes or made a decision favouring one alternative despite reasons favouring another.

Cognitive dissonance theory The theory that changes in attitudes can be motivated by an unpleasant state of tension caused by a disparity between a person's beliefs or attitudes and behaviour, especially beliefs or attitudes that are related to the person's self-esteem.

Cognitive illusion An error in judgement caused by a systematic way of thinking.

Cognitive learning This is the highest type of learning. An individual learns simply because he wants to learn. Purposive learning.

Cognitive learning theory Explanation of learning that focuses on mental processes.

Cognitive map A mental representation of an environment's spatial layout.

Cognitive neoassociation analysis The view that unpleasant experiences create negative affect, which in turn stimulates associations connected with anger and fear. Emotional and behavioural outcomes then depend, at least in part, on higher-order cognitive processing.

Cognitive neuropsychology A field of inquiry in which evidence of damage to certain areas of the brain and corresponding changes in behaviour are used to make inferences about underlying psychological functions.

Cognitive perspective The perspective on psychology that stresses human thought and the processes of knowing, such as attending, thinking, remembering, expecting, solving problems, fantasizing, and consciousness.

Cognitive processes Higher mental processes, such as perception, memory, language, problem solving, and abstract thinking.

Cognitive psychology The psychological science which studies cognition, the mental processes that are hypothesised to underlie behaviour. This covers a broad range of research domains, examining questions about the workings of memory, attention, perception, knowledge representation, reasoning, creativity and problem solving.

Cognitive reappraisal Any coping strategy in which one alters one's perception of the threat posed by a stressor to reduce stress.

Cognitive restructuring The process of replacing the client's maladaptive thoughts with more constructive ways of thinking.

Cognitive science The interdisciplinary field of study of the approach systems and processes that manipulate information.

Cognitive structures According to Piaget, mental representations or rules, such as schemata or concepts, that are used for understanding and dealing with the world and for thinking about and solving problems.

Cognitive tests Tests that measure an individual's learning in a specific subject area and quantify aspects of one's mental abilities.

Cognitive theory of aggression A fairly modern theory that suggests that aggression stems from complex interactions between cognition, affective states (emo-tion), and other additional components.

Cognitive therapy Cognitive therapy or cognitive behaviour therapy is a kind of psychotherapy used to treat depression, anxiety disorders, phobias, and other forms of mental disorder. It involves recognising distorted thinking and learning to replace it with more realistic substitute ideas. Its practitioners hold that much (though not all) clinical depression is associated with (although not necessarily caused by) irrational thoughts. Cognitive therapy is often used in conjunction with mood stabilizing medications to treat bipolar disorder. According to the U.S-based National Association of Cognitive-Behavioural Therapists:.

Cognitive-social approach to personality In personality studies, an approach to the study of behaviour which emphasizes the individual's expectations and hypotheses about the social world.

Cohabitation To live together as husband and wife, especially when not legally married.

Cohesiveness A "we feeling"; the extent to which members of a group are bound together, such as by attraction for one another.

Cohort effects The effects of being born and raised in a particular time or situation where all other members of your group has similar experiences that make your group unique from other groups.

Coimetrophobia Fear of cemeteries.

Coitophobia Fear of coitus.

Collaboration Professionals working cooperatively to provide educational services.

Collateral sprouts New branches grown on previously damaged axons, allowing some recovery of function.

Collective People engaged in common activities but having minimal direct interaction.

Collective effort model for achieving outcomes that they value.

Collective psychism The notion that living beings having a similar body share a group mind and collective consciousness and they influence each other and function together psychically.

Collective unconscious Collective unconscious is a term of analytical psychology, and was originally coined by Carl Jung. He distinguished the collective unconscious from the personal uncons-cious, which is particular to each human being. The collective unconscious refers to that part of a person's unconscious which is common to all human beings. It contains archetypes, which are forms or symbols that are manifested by all people in all cultures. Some have pointed out that this is essentially metaphysics since it is a hypothesis that can never be empirically confirmed or falsified; it was simply formulated by Jung as a model. The difference in their conceptualization of the unconscious is one of the more conspicuous differences between the psychologies founded by Jung and Freud.

Collectivism A cultural pattern in which people are considered to be fundamentally interdependent and obligations within one's family and immediate community are emphasized. Many of the societies of Latin America, and most of the cultures of Asia and Africa, are collectivist.

Colostrum The first fluid, rich in protein, secreted by the mother's mammary glands for several days just after birth of the young.

Colour blindness A hereditary condition that makes people unable to distinguish between colours.

Colour circle A simple way of illustrating how perceived hue and saturation depend upon the wavelength and spectral composition of a light and which can be used to predict the perceived colour of mixtures of monochromatic light.

Colour constancy The perceptual phenomenon by which the perceived colour of a surface remains constant despite variations in the spectral composition of the light that it reflects.

Colour disk A two-dimensional object allowing one to display two of the dimensions of colour: saturation and hue. Saturation is represented by radius (with achromatic colours at the center of the disk and fully saturated colours at the periphery), and hue is represented by angular position around the disk.

Colour mixing The perception of two or more lights of different wavelengths seen together as light of an intermediate wavelength.

Colour solid A three-dimensional object allowing one to display all three 'dimensions of colour: brightness, satura-tion, and hue. Brightness is represented by height (with black at the bottom and white at the top), saturation by radius (with achromatic colours at the center of the solid and fully saturated colours at the periphery), and hue by angular position around the solid.

Combat fatigue A traumatic neurosis characterized by the presence of somatic disturbances and anxiety reactions that have been brought on by extended exposure to combat. To layman's it has been called 'shell shock'.

Cometophobia Fear of comets.

Commitment The intent to continue a romantic relationship even in the face of difficulties.

Common sense As used in the discussion of artificial intelligence, the term refers to an understanding of what is relevant to a problem.

Communal relationship A relationship in which the participants expect and desire mutual responsiveness to each other's needs.

Community mental health movement A movement that advocates treating people with psychological problems in their own communities, providing outpatient treatment, and preventing psychological disorders.

Community psychology A form of treatment and education whose goal is to address psychological problems through an assessment of the sociocultural context in which they develop.

Comorbidity The experience of more than one disorder at the same time.

Companionate love A state of emotion (usually contrasted with romantic love) characterized by the affection we feel for those whose lives are deeply intertwined with our own.

Comparative method A research method in which one makes systematic comparisons among different species in order to gain insights into the function of a particular structure or behaviour, or the evolutionary origins of that structure or behaviour.

Comparative psychology A branch of psychology that studies the behaviours of a variety of organisms in an attempt to understand the adaptive and functional significance of the behaviours and their relation to evolution.

Compassionate love Warmth, trust, and tolerance of a person with whom one is romantically involved.

Compensation According to Alfred Adler, the process of striving to get rid of normal feelings of inferiority.

Compensatory education A program that is designed to prevent or remediate learning problems for students who are from lower socioeconomic status communities.

Compensatory preschool programs Programs designed to prepare disadvantaged children for entry into kindergarten and first grade.

Compensatory reaction An internally produced response through which the body seeks to reduce the effects of some external influence by producing a reaction opposite in its characteristics to those of the external influence. For example, the body produces an increase in pain sensitivity in response to the decrease in pain sensitivity caused by morphine, thereby canceling out morphine's reaction and so producing drug tolerance.

Competence That aspect of a theory which is the abstract knowledge which helps to describe some phenomena, as opposed to the performance theory. In language research a competence theory tries to capture generalizations about what language is and what it does.

Competition A striving or vying with others who share the same ecological niche for food, mates, and territory.

Complementarity The popularly supposed tendency, in a relationship between two people, for each to complete what is missing in the other. The questionable complementarity hypothesis proposes that people attract those whose needs are different, in ways that complement their own.

Complementary colour A colour that complements any colour evoked by a monochromatic light, such that when the two lights, e.g. Red-green, blue-yellow are mixed in equal proportions, the resulting colour is white.

Complementary colours Colours opposite each other on the colour circle; when additively mixed, they create the sensation of white light.

Complementary coloured after image A phenomenon which occurs when the eye adapts to one colour (e.g. Red) and then the gaze is shifted to an achromatic (e.g. Grey) region, an after image of the complementary colour (e.g. Green) is seen for a short time.

Completion items Fill-in-the-blank items on tests.

Complex cells A type of cell in the visual cortex that is sensitive to an input's orientation and so fires at its maximal rate only if the input is tilted appropriately. These cells are often sensitive to the direction of movement of a target.

Complex partial seizure disorder (CPSD) A kind of epilepsy that seems to make neurons within the amygdala hyperactive, leading sufferers to attach inappropriate emotional and motivational significance to objects, places, and events.

Complexity The range of wavelengths in light.

Compliance In group membership, the act of yielding to the will of others within the group. Conjunction fallacy in the method of anchoring and adjustment used in heuristics, the error of selecting a second statement without recognizing that everything that is both a and b must also be a.

Componential intelligence According to Sternberg, the mental mechanisms people use to plan and execute tasks; includes metacomponents, performance components, and knowledge acquisition components.

Compos mentis To be competent, or not to be legally insane or mentally deficient. In contrast to non compos mentis.

Compulsion A compulsion is an irrational need to perform some action, often despite negative consequences.

Compulsion neurosis A mental illness whose symptoms are either prohibitions and atonements (e.g., washing one's hands over and over) or symbolic substitute gratifications. As always for Freud, the problem's core is repressed sexual wishes—in this case the desire to masturbate. We now call this Obsessive-Compulsive Disorder.

Computer assisted tomography (CAT scan) An instrument for measuring and analyzing the brain's structure.

Computer-based instruction(CBA) Individualized instruction administered by a computer.

Computerized tomography (CT) A method for studying the brain that involves taking x-rays of the brain from different angles.

Concept A class or category that subsumes a number of individual instances. An important way of relating concepts is through propositions, which make some assertion that relates a subject (e.g., *chickens*) and a predicate (e.g., *lay eggs*).

Conception When a sperm and egg unite, resulting in an embryo or foetus.

Conceptual dependency (CD) A form of case grammar which aims to represent the key semantic concepts in language and to capture how these concepts depend upon each other.

Concordance The probability that a person who stands in a particular familial relationship to a patient (e.g., an identical twin) has the same disorder as the patient.

Concordance rate The percentage of both people in a pair having a certain trait or disorder.

Concordance research Research that studies the degree of similarity in traits expressed between twins. Twins are said to be concordant for a trait if either both or neither twin expresses it and discordant if only one twin expresses it.

Concrete operational stage The Theory of Cognitive development offered by Jean Piaget is often criticized by more recent psychologists who claim it underestimated the cognitive skills of younger children and overestimated the abilities of adolescents relative to adults. However, if the test of conservation he conducted is attempted - filling a glass of water and then transferring the water to another container of equal size and asking "which has more water, this or the other glass?", on children prior to the operational stage they normally will be unable to answer or will answer incorrectly due to, most likely, centration.

Concrete operations period In Piaget's theory, the period from ages six or seven to about eleven..At this time, the child has acquired mental operations that allow him to abstract some essential attributes of reality, such as number and substance, but these operations are as yet applicable only to concrete events and cannot be considered entirely in the abstract.

Concurrent validity "The extent to which an independent measure of an already measured trait shows similar results to the first/former test. ">The extent to which an independent measure of an already measured trait shows similar results to the first/former test.

Condensation The dream's tendency to combine several themes into one dream symbol. In this way the symbol can stand for several different thoughts, feelings, wishes, ideas.

Conditional probability The probability that one event will occur given that some other event has occurred.

Conditional response In classical conditi-oning, the response elicited by the CS.

Conditional stimulus In classical conditio-ning, a stimulus which, because of its repeated association with the UCS, eventually elicits a CR.

Conditioned (or Secondary) Reinforcer (or Punisher) A stimulus that acquires its rein-forcing (or punishing) properties through association with a primary reinforcer (or punisher). Sometimes referred to as a secondary reinforcer (or punisher).

Conditioned emotional response (CER) A technique in which a conditioned stimulus evokes fear, which in turn suppresses whatever other activities the animal is currently engaged in. For example, a rat will no longer press a lever for a food reward after several trials involving a light or tone that precedes an electrical shock.

Conditioned flavor-aversion learning A type of learning in which a substance is avoided because its flavor has been associated with illness.

Conditioned reinforcer An initially neutral stimulus that acquires reinforcing pro-perties through pairing with another stimulus that is already reinforcing.

Conditioned response (CR) A response elicited by some initially neutral stimulus, the conditioned stimulus (CS), as a result of pairings between that CS and an unconditioned stimulus (US). This CR is typically not identical with the uncon-ditioned response though it often is similar to it.

Conditioned stimulus (CS) In classical conditioning, the stimulus which comes to elicit a new response by virtue of pairings with the unconditioned stimulus.

Conditioning The process of learning new behaviours or responses as a result of their consequences.

Conditions of worth Conditions that others place on us for receiving their positive regard.

Conduct disorders Socioemotional and behavioural disorders indicated in indivi-duals who, for example, are chroni-cally disobedient or disruptive.

Conduction aphasia An inability to remember words that are heard, although they usually can be understood and responded to appropriately. This disability is caused by damage to Wernicke's and Broca's areas.

Cones A type of retinal receptor that is active in daylight (photopic) conditions. The three different types of cone found in the human retina provide the basis for colour vision.

Confabulation Sincere but false recollections, usually produced when one encounters a gap in the memory record and (unwittingly) tries to fill this gap.

Confederate Accomplice of an experimenter who, in dealing with the real participants in an experiment, acts as if he or she is also a participant.

Confidence interval An interval around a sample mean within which the population mean is likely to fall. In common practice, the largest value of the interval is 2 standard errors above the mean and the smallest value is 2 standard errors below it.

Confidentiality The protection of the anonymity of a participant in a psychological study through the correct storage and published use of their data.

Confirmation bias The tendency to look for and accept evidence that supports what one wants to believe and to ignore or reject evidence that refutes those beliefs.

Conflict An action taken to block or interfere with others' interests, because of the perception that ones opponent is doing the same, or that each others actions are incompatible.

Conflict-resolution strategy The expression of heuristics in a uniform way as sets of 'if...then' rules or ways of deciding what to do when the conditions for more than one production are satisfied.

Conformity Conformity is the act of maintaining a certain degree of similarity (in clothing, manners, behaviours, etc.) To those in your general social circles or to those in authority. Usually, conformity implies a tendency to submit to others in thought and behaviour other than simply clothing choice. It can also refer to.

Confound Any variable that is not part of a research study but still has an effect on the research results.

Confounding factor In statistics, a confounding factor is a factor which is the common cause of two things that may falsely appear to be in a causal relationship. It is the cause of a spurious relationship.

Confounding variable A stimulus other than the variable an experimenter explicitly introduces into a research setting that affects a participant's behaviour.

Congenital defects Birth defects existing as such at birth.

Congruence According to Carl Rogers, the accurate match between self-concept and reality.

Conjugate movement The cooperative movement of the eyes, which ensures that the image of an object falls on identical portions of both retinas.

Conjunction fallacy In the representative heuristic, an error of judgement where the second of two statements is selected because a person fails to recognize that everything that is both a and b must also be a.

Conjunction of features In a visual search procedure, a target that is composed of several different features (i.e., a red X as opposed to the feature red or the feature diagonal). Search times required to find these kinds of targets are longer and increase with the number of distractors that are displayed.

Connectionism In cognitive psychology, the major formal language for the description of sub-symbolic learning.

Connectionist models A model of how information in memory is re-trieved that relies on distributed representations. In a distributed representation, a concept is conveyed by a pattern of activation across an entire network, rather than by the activation of a single node. In such models, processing depends on having just the right links between concepts, at just the right strengths.

Conscience Conscience is generally thought of as a moral faculty, sense, or feeling that impels individuals to believe that particular activities are morally right or wrong.

Conscience One of two sectors that comprise the superego. Basically, an internalized critical parent. The conscience turns our innate destructiveness back upon ourselves and thereby safeguards the cohesion of society. At the same time, renunciations of drive strengthen the conscience. Unlike the ego ideal's internalized standards, those of the conscience deal mainly with what is "bad." Ultimately, it is fear of castration.

A severe conscience comes about by 1. frustration of a drive, which releases aggression needed to build the conscience, and

2. being loved, which directs innate aggression back to the person who has it.

Conscientiousness Tendency to set high goals, to accomplish work successfully, and to behave dutifully and morally.

Conscious A quality of the mental life of the ego. The ego uses consciousness to manage reality, which includes allowing pleasurable id gratifications that don't collide with everyday demands or provoke the "thou shalt not"s of the superego.

Consciousness Consciousness is a quality of the mind generally regarded to comprise qualities such as subjectivity, self-awareness, sentience, sapience, and the ability to perceive the relationship between oneself and one's environment. Philosophers divide consciousness into phenomenal consciousness which is experience itself and access consciousness which is the processing of the things in experience (Block 2004).

Consecotaleophobia Fear of chopsticks.

Consensual behaviour Behaviour that is shared by many people; behaviour that is similar from one person to the next. To the extent that people engage in the same behaviour, their behaviour is consensual.

Consensual validation In group membership, the process by which the group affirms and supports the views of a particular individual.

Consent One of the components of the 'Ethical Principles for Conducting Research with Human Participants' drawn up by the British Psychology Society, which underlines the participant's need to consent to a particular study and be fully informed about the nature of the work.

A clinical practitioner who specializes in helping and advising adults in troubled and dysfunctional relationships, or who are struggling with social problems such as drug or alcohol addiction through interpersonal procedures such as interviewing and tests.

Consequence A condition that follows a behaviour and affects the frequency of future behaviour.

Conservation According to Piaget, the understanding that physical properties do not change when nothing is added or taken away, even though appearances may change.

Conservation of number In Piaget's theory, the understanding that the number of objects in a group remains constant despite any changes in their spatial arrangement (e.g., a child at age six realizes that there is the same number of objects in a row of six closely spaced bottles as in a row of six bottles spaced far apart).

Conservation of quantity In Piaget's theory, the understanding that the quantity of a substance remains unchanged despite a visible change in appearance (thus in liquid conservation, the realization that the amount of liquid remains the same when poured from a tall, thin beaker into a short, wide jar).

Conservation task A type of task, described by jean piaget, where superficial changes in an object or group of objects do not necessarily alter the properties of those objects.

Consistency paradox The observation that personality ratings across time and among different observers are consistent, while behaviour ratings across situations are not consistent.

Consistency The extent to which a person's behaviour is consistent across time.

Consolidation The process by which information in short-term memory is transfered to long-term memory, presu-mably because of physical changes that occur in neurons in the brain.

Consolidation failure The failure to store information in memory.

Conspecific In sociobiology, a name for another animal of the same species.

Constant Any variable that remains the same throughout a study.

Construct Any variable that can not be directly observed but rather is measured through indirect methods. (Examples: intelligence, motivation).

Construct validity The construct is a collection of related behaviours that are associated in a meaningful way. A judgemental-empirical approach to testing a previously hypothesized relationship.

Constructive play A stage of play amongst children when they develop skills by using objects in non-pretend ways, such as doing a jigsaw or pouring sand or water from one container into another.

Constructivism Theories of cognitive development that emphasize the active role of learners in building their own understanding of reality.

Constructivist theories of learning State that learners must individually discover and transform complex information, checking new information against old rules and revising them when they no longer work.

Consummate love The ideal form of love in adulthood that involves three components. Passion, intimacy, and commitment.

Contact comfort Comfort derived from an infant's physical contact with the mother or caregiver.

Contact hypothesis A hyposthesis stating that prejudice declines when people in an ingroup become more familiar with the customs, norms, food, music, and attitudes of people in an outgroup.

Contagion In the study of crowd behaviour, the spread of feeling or a mood through a crowd which leads them to behave in particular, concerted and often violent ways.

Contamination When the comparison group is in some way affected by, or affects, the treatment group, causing an increase of efforts. Also known as compensatory rivalry or the John Henry effect.

Content morphemes Morphemes that carry the main burden of meaning (e.g., *strange*). This is in contrast to function morphemes that add details to the meaning but also serve various grammatical purposes (e.g., the suffixes *s* and *er*, the connecting words *and*, *or*, *if*, and so on).

Content validity The extent to which the scope of a test is broad enough to encompass all aspects of the definiton of what you're measuring—the definition contains the concept.

Content word A noun, verb, adjective, or adverb that conveys meaning. See also *function word*.

Context In encoding specificity experiments, intrinsic context refers to various features that are an integral part of a target stimulus, whilst extrinsic context represents those features that are present when the target is encountered.

Context dependent memory The theory that information learned in a particular situation or place is better remembered when in that same situation or place.

Context of discovery The initial phase of research, in which observations, beliefs, information, and general knowledge lead to a new idea or a different way of thinking about some phenomenon.

Context of justification The research phase in which evidence is brought to bear on hypotheses.

Contextual distinctiveness The assumption that the serial position effect can be altered by the context and the distinctiveness of the experience being recalled.

Contextual intelligence According to Sternberg, intelligence that reflects behaviours that were subject to natural selection: *adaptation*-fitting into the environment by developing useful skills and behaviours; *selection*-finding a niche in the environment; and *shaping*-changing the environment.

Contiguity The togetherness in time and space of two events, which is sometimes regarded as the condition that leads to association.

Contiguity learning, classical conditioning Most primitive learning. Learning by association (experiences occurring together in space or time go together) through the principles of contiguity (nearness), similarity and contrast.

Contingency A relation between two events in which one is dependent upon another. If the contingency is greater than zero, then the probability of event *A* will be greater when event *B* is present than when it is absent.

Contingency management A form of behaviour therapy in which the environment is structured such that certain behaviours are reliably followed by well-defined consequences.

Contingency model of leadership The theory that leadership effectiveness depends both on whether leaders are task oriented or relationship oriented, and on the degree to which they have situational control.

Contingent praise Praise that is effective because it refers directly to specific task performances.

Continuing education classes Taken by adults to expand their knowledge or skills for personal or work-related development.

Continuity The tendency to perceive interrupted lines and patterns as being continuous by filling in gaps.

Continuity versus discontinuity Debate examines the question of whether development is solely and evenly continuous, or whether it is marked by age-specific periods.

Continuous reinforcement A reinforcement schedule in which reinforcement happens every time a particular response occurs.

Continuous theory of development Theory based on the belief that human development progresses smoothly and gradually from infancy to adulthood.

Contralateral Residing in the side of the body opposite the reference point.

Contralateral control The pattern in which movements on the right side of the body are controlled by the left half of the brain, while movements on the left side of the body are controlled by the right half of the brain. Contralateral control is seen in nearly all vertebrate nervous systems.

Contrast effect The tendency to perceive stimuli that differ from expectations as being even more different than they really are.

Contreltophobia Fear of sexual abuse.

Control group A group of subjects in an experiment that receives the same treatment and is treated exactly like the experimental group, except with respect to the independent variable.

Control procedures Consistent procedures for giving instructions, scoring responses, and holding all other variables constant except those being systematically varied.

Controlled processes Processes that require attention; it is often difficult to carry out more than one controlled process at a time.

Conventional level Kohlberg's second level of moral development, in which people realize that society has instituted moral rules to maintain order and to serve the best interests of its citizenry.

Conventional level of morality Stages 3 and 4 in Kohlberg's model of moral development, in which individuals make moral judgments in consideration of others.

Conventional morality Characterized by conformity, helping others, obeying the law, and keeping order.

Convergence In the absence of a more specific context, convergence denotes the approach toward a definite value, as time goes on; or to a definite point, a common view or opinion, or toward a fixed or equilibrium state.

Convergent and divergent production Convergent and divergent production are the two types of human response to a set problem that were identified by J. P. Guilford. Convergent production is the deductive generation of the best single answer to a set problem, usually where there is a compelling inference. For example, find answers to the question What is the sum of the internal angles of a triangle?.

Convergent evolution An increase in similarity between species in response to the same selection pressure brought about by the problems of living in a particular niche.

Convergent thinking A style of thinking in which a person narrows down a list of possibilities to arrive at a single right answer.

Conversion A defence mechanism that involves converting an intrapsychic conflict into a physical form, such as blindness, deafness, paralysis, or numbness.

Conversion disorder A somatoform disorder involving the actual loss of bodily function, such as blindness, paralysis, and numbness, due to excessive anxiety.

Convolutions The wrinkles visible in the cortex that allow the enormous surface area of the human cortex to be stuffed into the relatively small space of the skull.

Convulsive disorders Forms of epilepsy.

Cooperative learning Adult-supervised education that relies on peers' interacting, sharing, planning, and supporting each other.

Cooperative play Play in which children join together to achieve a common goal.

Cooperative scripts A study method in which students work in pairs and take turns orally summarizing sections of material to be learned.

Coordinated reference system In environmental psychology, the system of abstract geometric patterns which children acquire by the end of piaget's concrete operations stage.

Coping Efforts to reduce stress.

Coping resources Those resources both within the individual and in the environment around them which enable them to deal with stressful situations.

Coping response A response that permits an animal to escape, avoid, or minimize the stressful (harmful or painful) effects of an aversive stimulus.

Coping strategy A plan of action that a person follows to reduce the perceived level of stress, either in anticipation of a stressor or in response to its occurrence.

Coprastasophobia Fear of constipation.

Coprophobia Fear of feces.

Cornea The transparent outer membrane of the eye.

Corpus callosum The connecting nerve bundle that bridges the two halves of the brain and provides continuity between the right and left hemispheres as electrical impulses pass to each side.

Corpus luteum During the early stages of pregnancy, a mass of yellow tissue formed in the ovary by a ruptured graafian follicle that has discharged its ovum.

If the ovum is fertilized, this tissue secretes the hormone progesterone, needed to maintain pregnancy.

Correctional mechanism In a regulatory process, the mechanism that is capable of restoring the system variable to the set point.

Corrective instruction Educational activities that are given to students who initially fail to master an objective; designed to increase the number of students who master educational objectives.

Correlated sample Sample data that is related to each other.

Correlation The degree to which two or more variables a related to each other. A correlation refers to the direction that the variables move and does not necessarily represent cause and effect. (Example: height and weight are correlated. As one increases, the other tends to increase as well).

Correlation coefficient (r) A number that expresses both the size and the direction of a correlation, varying from +1.00 (perfect positive correlation) through 0.00 (absence of any correlation) to -1.00 (perfect negative correlation).

Correlational methods Research methodologies that determine to what extent two variables, traits, or attributes are related.

Correlational research method A research method that provides information about the relationship between variables. It is also called a descriptive research method.

Correlational studies Studies in which the investigator is seeking to observe the relationship among variables that were in place prior to the study (as opposed to factors that the investigator creates or manipulates).

Correspondence problem In a moving display, the difficulty in determining which aspects of the display now visible correspond to which aspects of the display visible a moment ago.

Correspondent inference An inference that the action of an actor corresponds to, or is indicative of, a stable personal characteristic.

Correspondent inference theory A theory holding that we make inferences about a person when his or her actions are freely chosen, are unexpected, and result in a small number of desirable effects.

Cortex The higher areas of the brain, which are responsible for thinking and planning.

Cortical plasticity The ability of the cerebral cortex to change as a function of learning or damage, such as the reacquisition of language function following damage to the language hemisphere.

Corticosteroids Hormones released by the adrenal cortex in response to stress.

Coulrophobia Fear of clowns.

Counselling psychologist A clinical practitioner who specializes in helping and advising adults in troubled and dysfunctional relationships, or who are struggling with social problems such as drug or alcohol addiction through interpersonal procedures such as interviewing and tests.

Counterbalancing A systematic variation of conditions in an experiment, such as the order of presentation of stimuli, so that different subjects encounter them in different orders; prevents confounding of independent variables with time-dependent processes such as habituation or fatigue.

Counterconditioning A procedure for weakening a classically conditioned response (CR) by connecting the stimuli that presently evoke it to a new response that is incompatible with the CR.

Counterfactual thinking The tendency to imagine alternative events or outcomes that might have occurred but did not.

Counterphobia The preference by a phobic for fearful situations.

Countertransference Circumstances in which a psychoanalyst develops personal feelings about a client because of perceived similarity of the client to significant people in the therapist's life.

Counterwish In a dream, an element that appears to contradict a wish but actually fulfills it. The frustration dreams of a masochist, for instance, who gets off on being frustrated.

Couples therapy A type of therapy in which a therapist helps couples identify and resolve conflicts.

Covariation principle A principle of attribution theory stating that for something to be the cause of a particular behaviour, it must be present when the behaviour occurs and absent when it does not occur.

Covert sensitization A method used by behaviour therapists in which a client imagines the aversive consequences of his or her inappropriate behaviour.

Cranial nerves The twelve pairs of nerves that enter and exit directly from the hindbrain. These nerves control movements of the head and neck, carry sensations from them including vision, olfaction, and audition, and regulate the various glandular secretions in the head.

Creative intelligence The form of intelligence alleged by some authors as essential for devising new ideas or new strategies. Often contrasted with analytic intelligence or practical intelligence.

Creativity The ability to generate ideas or products that are both novel and appropriate to the circumstances.

Credibility Believability. A credible communicator is perceived as both expert and trustworthy.

Cremnophobia Fear of precipices.

Creole A pidgin language which has become the native tongue of the children growing up in a pidgin speech community.

Criminal behaviour profiling A criminal investigation technique in which crime scenes and additional evidence are analyzed in order to discern patterns in the offenders behaviour, with which a behavioural and physical description of the offender can be created. Has also been called: "profiling", "offender profiling", "investigative profiling".

Criminal type A category of individuals whom repeatedly engage in criminal and/ or antisocial behaviour. They apparently have a constitutional tendency towards behaving in that direction.

Criminology The scientific study of crime, criminals, and penology. This science often considers both social and psychological aspects of criminality.

Criterion An independent measure of a variable being assessed. For example, college grades are the criterion measure for scores on the Scholastic Aptitude Test.

Criterion groups Groups whose test performance sets the validity criterion for certain tests (e.g., the Minnesota Multiphasic Personality Inventory, MMPI, which uses several psychiatric criterion groups to define most of its subscales).

Criterion referenced evaluations Assessments that rate how thoroughly students have mastered specific skills or areas of knowledge.

Criterion validity The degree to which test scores indicate a result on a specific measure that is consistent with some other criterion of the characteristic being assessed; also known as predictive validity.

Critical period A period in the development of an organism when it is particularly sensitive to certain environmental influences. Outside of this period, the same environmental influences have little effect (e.g., the period during which a duckling can be imprinted). After embryonic development, this phenomenon is rarely all-or-none. As a result, most developmental psychologists prefer the term *sensitive period.*

Critical ratio A score, usually a *z*-score, that determines whether an investigator will accept or reject the null hypothesis. If a test score exceeds the critical ratio, the null hypothesis is rejected.

Critical thinking Ability to make rational decisions about what to do or what to believe.

Critical value The value of a statistic required in order to consider the results significant.

Cross age tutoring Peer tutoring between an older and a younger student.

Cross cultural research Research designed to compare and contrast people of different cultures.

Cross-cultural psychology The branch of psychology that studies the effects of culture on behaviour.

Cross gender behaviours Behaviours stereotypical of the opposite sex.

Cross race identification bias The tendency for people to have difficulty identifying members of a race other than their own.

Cross-sectional design A research method in which groups of participants of different chronological ages are observed and compared at a given time.

Cross sectional study A study in which a number of different-age individuals with the same trait or characteristic of interest are studied at a single time.

Cross sequential study A research study that examines the effects of development (maturation) by combining longitudinal and cross sectional studies.

Crowding The psychological and psychological response to the belief that there are too many people in a specified area.

Crowning The point during labor when the baby's head can be seen at the vaginal orifice.

Cryophobia Fear of extreme cold, ice or frost.

Crystallized intelligence The facet of intelligence involving the knowledge a person has already acquired and the ability to access that knowledge; measures by vocabulary, arithmetic, and general information tests.

Crystallophobia Fear of crystals or glass.

CT scan (Computerized Tomography) A technique for examining brain structure in living humans by constructing a composite X-ray picture based on views from all different angles. Also called CAT (Computerized Axial Tomography) scan.

CT scanner A device that uses a special X-ray machine and a computer to produce images of the brain that appear as slices taken parallel to the top of the skull.

Cue A particular feature of the stimulus that forms the basis for further processing, for example a depth cue.

Culminating phase A phase of early adulthood that ranges from ages 33 to 45.

Cult (also called new religious movement) A group typically characterized by 1. Distinctive ritual and beliefs related to its devotion to a god or a person,

2. Isolation from the surrounding "evil" culture, and

3. A charismatic leader. (a sect, by contrast, is a spinoff from a major religion.).

Cultivation The process by which the mass media (particularly television) construct a version of social reality for the public.

Cultural anthropology A branch of anthropology that compares the similarities and differences among human cultures.

Cultural display rules Learned but deeply ingrained conventions that govern what facial expressions of emotion may or may not be shown in what contexts.

Cultural evolution The adaptive change of a culture to recurrent environmental pressures.

Cultural perspective The psychological perspective that focuses on cross-cultural differences in the causes and consequences of behaviour.

Culture The total lifestyle of a people from a particular social grouping, including all the ideas, symbols, preferences, and material objects that they share.

Culture bound syndromes Highly unusual mental disorders, similar in nature to nonpsychotic mental disorders, that appear to be specific to only one or a few cultures.

Culture fair IQ tests Tests that are fair for all members in a culture.

Culture fairness of a test The extent to which test performance does not depend upon information or skills provided by one culture but not another.

Culture free IQ tests Tests without cultural content.

Culture, two purposes of To protect us against nature, and to regulate our affairs with each other. To adapt to life in a culture, one must repress one's drives (repress one's sexuality and aggression, displacing the second onto suitable targets outside the culture and sublimating the first into other achievements). "Man is a savage beast," and to repeat a quotation from Plautus which Freud liked, "Homo homini lupus" ("Man is a wolf to man"—written, presumably, by a wolf). So Eros and Ananke (Love and Necessity) are the parents of civilization, and social restrictions on sexuality are unavoidable.

Were it not for our need to live with one another, we could allow our drives free play and not be neurotic.

Note: this term, kultur in German, is often mistranslated civilization, as in Civilization and its Discontents ("Discontents" per Freud should be translated "discomfort" or "malaise").

Cumulative recorder A mechanical device connected to an operant chamber for the purpose of recording operant responses as they occur in time.

Cupboard theory A hypothesis about the infant's attachment to the primary caregiver; according to this theory, the attachment is motivated largely by the fact that the mother is a source of nourishment (whether through breast or bottle).

Cutaneous senses The skin senses that register sensations of pressure, warmth, and· cold.

Cutoff score Score designated as the minimum necessary to demonstrate mastery of a subject.

Cutting off reflected failure Actively disidentifying with and distancing oneself from the failures or negative evaluations of others.

Cyberphobia Fear of computers or working on a computer.

Cycle of family violence The transmission of domestic violence across generations.

Cyclophobia Fear of bicycles.

Cyclothymic Oscillation, reversing, frequently changing, .in a cyclic fashion.

Cymophobia or kymophobia Fear of waves or wave like motions.

Cynophobia Fear of dogs or rabies.

Cypridophobia or cypriphobia or cypriano-phobia or cyprinophobia Fear of prostitutes or venereal disease.

Cytochrome oxidase blobs An anatomical (and possibly functional) feature of the visual cortex revealed by selectively staining with cytochrome oxidase.

D

Dactyloscopy This is the scientific analysis of fingerprints. Fingerprint experts have been involved with law enforcement for almost a hundred years. Recent advances in dactyloscopy have continued to make fingerprint analysis of key importance (i.e., new methods pull fingerprints off underwater surfaces, skin, etc.).

Dark adaptation The process by which the eye becomes capable of distinguishing dimly illuminated objects after going from a bright region to a dark one.

Darwinian fitness In the theory of evolution, the ability to reproduce successfully and leave descendants, a fact which is dependant upon particular attributes rather than physical health.

Date rape Unwanted sexual violation by a social acquaintance in the context of a consensual dating situation.

Daytime sleepiness The experience of excessive sleepiness during daytime activities; the major complaint of patients evaluated at sleep disorder centers.

Deadline technique A technique used in many areas, in which the target is told that they have only a limited time to accept an offer. Often used in criminal interrogations (i.e. "if you confess now the crown attorney will go easy on you").

Death The permanent cessation of all life functions.

Death education Provides people with information on dying, legal issues, and various practical matters.

Death instinct Should actually be translated "death drive." A fundamental tendency for life to seek the perfect equilibrium of entropic non-activity—in other words, dissolution. "The goal of life is death.".

Death qualification A jury-selection procedure used in capital cases that permits judges to exclude prospective jurors who say they would not vote for the death penalty.

Debriefing A disclosure, made to participants after research procedures are completed, in which the researcher explains the purpose of the research, attempts to resolve any negative feelings, and emphasizes the scientific contribution made by the participants' involvement.

Decay theory A theory of forgetting in which memory traces erode largely through the passage of time (presumably because of some metabolic events unfolding as time passes).

Decentration The ability to focus simultaneously on several aspects of a problem.

Deception A research technique that provides false information to persons participating in a study.

Decibels The logarithmic units used to describe sound intensity (or amplitude).

Decidophobia Fear of making decisions.

Deciduous teeth Baby teeth; teeth that fall out at a certain stage of growth.

Decision aversion The tendency to avoid decision making; the tougher the decision, the greater the likelihood of decision aversion.

Decision Making the decision to commit to a relationship with another person.

Decision making The process of forming probability estimates of events and using them to choose between different courses of action.

Declarative knowledge Knowing "that" (e.g., knowing someone's name) as contrasted with procedural knowledge, which is knowing "how" (e.g., knowing how to ride a bicycle).

Declarative memory Declarative memory is the aspect of memory that stores facts and events. It applies to standard textbook learning and knowledge. It is based on pairing the stimulus and the correct response. For example, the question "What is the capital of Sierra Leone?" is paired with the answer "Freetown". The name declarative comes from the fact that we can explicitly "ask" our brain to make a connection between a pair of stimuli. Declarative memory is subject to forgetting and requires repetition to last for years. Declarative memories are best established by using active recall combined with mnemonic techniques and spaced repetition.

Decussation The crossing of nerve fibres from one side of the brain to the other.

Deductive reasoning In traditional Aristotelian logic, deductive reasoning is inference in which the conclusion is of lesser or equal generality than the premises, as opposed to inductive reasoning, where the conclusion is of greater generality than the premises. Other theories of logic define deductive reasoning as inference in which the conclusion is just as certain as the premises, as opposed to inductive reasoning, where the conclusion can have less certainty than the premises. In both approaches, the conclusion of a deductive inference is necessitated by the premises: the premises can't be true while the conclusion is false. (In Aristotelian logic, the premises in inductive reasoning can also be related in this way to the conclusion.).

Deep processing The analysis of the complex characteristics of a stimulus, such as its meaning or its relationship to other stimuli.

Deep structure The essential meaning of a sentence, without regard to the grammatical features (surface structure) of the sentence that are needed to express it in words. See also *surface structure.*

Defecaloesiophobia Fear of painful bowels movements.

Defence mechanism A maneuver employed by the ego to protect itself against anxiety raised by intolerable impulses. All involve some degree of repression of the unacceptable impulse into uncons-ciousness. Examples include denial, idealization, splitting (e.g., both loving and hating someone but keeping both emotions entirely separate), reaction formation (e.g., becoming a Scoutmaster to prove to yourself that you don't hate children), undoing (basically, trying to repair an action for which you feel guilty), and an American favourite, intellectualization ("You ask me how I feel, and it seems to me that the relevancy of the issue has more to do with my sublimated urge to..." blah). Freud's daughter Anna discovered some more, perhaps because being analyzed by your father makes them so necessary to have handy.

Note: the English term defence is actually farther from what Freud wrote in German: abwehr means to parry or to ward off, not to defend. Deflection and to deflect are more accurate terms than defence and to defend.

Defensible space In environmental psychology, the individual's personalization of the particular area they inhabit and are familiar with and which they feel safe in. Deficiency motives in self-actualization, those motives which reduce the personal needs lower in the hierarchy, such as reducing hunger and thirst and the need for obtaining security.

Defensive wounds Wounds commonly found on a victims arms, hands, or fingers sustained when the victim was trying to defend self from an assault.

Deferred imitation A child's ability to imitate the actions he or she has observed others perform. Piaget believed deferred imitation to result from the child's increasing ability to form mental representations of behaviour performed by others.

Deferred imitation A pattern of imitation first observed in children late in the second year of life in which the child mimics an action observed some time in the past.

Deficiency needs Basic requirements for physical and psychological well-being as identified by Maslow.

Definition (of a word) A set of necessary and sufficient features shared by all members of a category that are the criteria for membership in that category.

Definitional theory of meaning The theory that our mental representation of word meaning is made up of a small number of simpler concepts. The representation of *bachelor*, for example, is made up of "adult," "unmarried," and "male.".

Degrees of freedom The term "degrees of freedom" is used in three different branches of science: in physics and physical chemistry, in mechanical and aerospace engineering, and in statistics. The three usages are linked historically and through the underlying mathematics through the concept of dimensionality, but they are not identical.

Deindividuation Loss of self-awareness and evaluation apprehension; occurs in group situations that foster responsiveness to group norms, good or bad.

Deinstitutionalization A movement intended to obtain better and less expensive care for chronically mentally ill patients in their own communities rather than at large, centralized hospitals.

Deipnophobia Fear of dining or dinner conversations.

Déjà raconté The patient's illusion of already having said or done something in the therapist's presence when this did not really occur. The patient indeed had the impulse, but didn't act on it. Often happens at payment time.

Delay of gratification The postponement of immediate satisfaction in order to achieve a more important reward later on, a process that plays an important role in some behavioural-cognitive approaches to personality.

Delayed inhibition The temporal equivalent of lateral inhibition, where responses are subtracted from a delayed copy of themselves so that unchanging stimuli are filtered out and only temporal changes are transmitted.

Deliberate practice A form of acquiring a skill under close supervision, in which specific aspects of the skill are trained, as opposed to general play in which there is no such deliberate plan.

Delinquency A type of reckless behaviour, particularly among adolescents, and sometimes influenced by hormonal changes, which often involves crime, underage drinking and sexual activity, and illegal drug use.

Delivery Expelling the baby and placenta from the vagina.

Delta activity The rhythmical activity of the electroencephalogram, having a frequency of less than 3.5 Hz, indicating deep (slow-wave) sleep.

Delta rule A powerful form of learning, also known as gradient descent learning, which suggests that following a particular action, the difference between the planned outcome and the actual outcome (the error) is calculated, and the system adjusted to reduce the error next time.

Delta waves The type of brain waves present when a person is deeply asleep.

Delusion A delusion is commonly defined as a false belief, and is used in everyday language to describe a belief that is either false, fanciful or derived from deception. In psychiatry, the definition is necessarily more precise and implies that the belief is pathological (the result of an illness or illness process).

Delusions of control The false belief that one's thoughts and actions are being controlled by other people or forces.

Delusions of grandeur The false belief that one is famous, powerful, or important.

Delusions of persecution The false belief that other people are plotting against one; symptom of schizophrenia.

Demand characteristics Cues in an experimental setting that influence the participants' perception of what is expected of them and that systematically influence their behaviour within that setting.

Dementia Dementia (from Latin demens) is progressive decline in cognitive function due to damage or disease in the brain beyond what might be expected from normal aging. Particularly affected areas may be memory, attention, language and problem solving, although particularly in the later stages of the condition, affected persons may be disoriented in time (not knowing what day, week, month or year it is), place (not knowing where they are) and person (not knowing who they are). Symptoms of dementia can be classified as either reversible or irreversible depending upon the etiology of the disease, although dementia, by definition, is irreversible and will eventually result in death. Dementia is a non-specific term that encompasses many disease processes just as fever is attributable to many etiologies.

Dementia with Lewy bodies Dementia with Lewy bodies is the second most frequent cause of hospitalization for dementia, after Alzheimer's disease. Current estimates are that about 60 to 75% of diagnosed dementias are of the Alzheimer's and mixed (Alzheimer's and vascular dementia) type, 10 to 15% are Lewy Bodies type, with the remaining types being of an entire spectrum of dementias including frontotemporal, Pick's disease, alcoholic dementia, pure vascular dementia, etc.

Dementophobia Fear of insanity.

Demonophobia or Daemonophobia Fear of demons.

Demophobia Fear of crowds. (Agoraphobia).

Dendrite In biology, a dendrite is a slender, typically branched projection of a nerve cell, or "neuron," which conducts the electrical stimulation received from other cells to the body or soma of the cell from which it projects. This stimulation arrives through synapses, which typically are located near the tips of the dendrites and away from the soma.

Dendritic trees Highly branched fibres extending from neurons.

Dendrophobia Fear of trees.

Denial One of Elisabeth Kubler-Ross's stages of dying, in which the dying refuse to acknowledge their inevitable death, perhaps believing a mistake has been made.

Density-intensity effect hypothesis In environmental psychology, the correlation between varying levels of social density and the individual's positive and/or negative reaction to them.

Dentophobia Fear of dentists.

Dependent variable A feature of a particular phenomenon which varies according to the theory because it depends directly on how a given situation is manipulated.

Depersonalization An offenders attempt to eliminate the identity of the victim so that they do not represent or resemble the person whom has caused their psychological distress. Actions can range from covering the victims face with a towel or blanket up towards extreme battery.

Depolarization A drop of the membrane potential of a neuron from its resting potential. The basis of neural excitation.

Depression A catch-all term used to describe any of several complex, and sometimes extreme psychological states where the emotions are affected, either by internal factors such as by low spirts and low self-esteem or external factors, such as traumatic events in someone's life.

Depressive explanatory style A habitual tendency to attribute negative events to internal, stable, and global causes, and positive events to external, unstable, and specific causes.

Depressive realism The tendency of mildly depressed people to make accurate rather than self-serving judgments, attributions, and predictions.

Depth cues Sources of information that signal the distance from the observer to the distal stimulus. Some depth cues are present in a single retinal image (the pictorial cues), some require a comparison of the information received from the two eyes (binocular cues), some involve the pattern of motion in the retinal image (parallax and optic flow), and some arise from the positions of the eyes in viewing (e.g., convergence angle).

Depth perception The ability to understand the relationship among objects in space. Human beings develop a sense of depth perception as infants.

Depth-of-processing approach An approach to memory that stresses the nature of encoding at the time of acquisition. It argues that deeper levels of processing (for example, attending to a word's meaning) lead to better retention and retrieval than shallower levels of processing (for example, attending to the word's sound). Thus, maintenance rehearsal leads to much poorer retrieval than elaborative rehearsal.

Derived feature In sociobiology, a feature whose possession is unique to the group of species descended from the single (originator) population.

Derived scores Values computed from raw scores that relate students1 performances to those of a norming group; examples are percentiles and grade equivalents.

Dermatophobia Fear of skin lesions.

Dermatosiophobia or dermatophobia or dermatopathophobia Fear of skin disease.

Descriptive hierarchy The idea that a stimulus is represented internally at several different levels of detail, with simple features like edges at the lower levels being combined into more complex features, like corners, at the higher levels.

Descriptive Research Study aimed at identifying and gathering detailed information about something of interest.

Descriptive statistics The branch of statistics that focuses on describing in numerical format what is happening now within a population. Descriptive statistics require that all subjects in the population (the entire class, all males in a school, all professors) be tested.

Design out In architectural planning, the taking into account of special needs or social factors affecting a particular area or community, such as housing estates, and making them less vulnerable to crime and vandalism by redesigning their layout.

Despair A loss of hope; Erik Erikson believed that those in late adulthood struggled with the fear that there is too little time to begin a new life course.

Detector In a regulatory process, a mechanism that signals when the system variable deviates from its set point.

Determinism The doctrine that all events—physical, behavioural, and mental—are determined by specific causal factors that are potentially knowable.

Deuteranopia A form of hereditary anomalous colour vision; caused by defective "green" cones in the retina.

Development The series of age-related changes that occurs over the course of a person's life span.

Developmental age The chronological age at which most children show a particular level of physical or mental development.

Developmental approach An approach to the study of intelligence that studies the way infants and children learn to perceive, manipulate, and think about the world.

Developmental norms The median ages at which children develop specific behaviours and abilities.

Developmental psycholinguistics The study of language and its development in children, as seen from a psychological point of view.

Developmental psychologists Psychologists who seek to understand, describe, and explore how behaviour and mental processes change over the course of a lifetime.

Developmental psychology The branch of psychology concerned with interaction between physical and psychological processes and with stages of growth from conception throughout the entire life span.

Developmentalists Researchers who study human development.

Developmentally appropriate education Instruction felt to be adapted to the current developmental status of children (rather than their age alone).

Developmentally disabled Persons who are slowed or delayed in development or progress, especially because of subnormal intellectual functioning and social skills.

Deviance Behaviour, ideas, and attributes that are responded to negatively by others. Going against the social norm, whether it is criminal (murder/rape), social (wearing red to a funeral), or physical (facial tattoo's, physical abnormality). Deviance can change depending on the setting of the act/attribute (cursing with friends vs. in church), the age of the actor (baby soiling cloths vs. teenager), etc.

Deviant One who acts defiantly, have socially deviant thoughts, or has socially abnormal attributes. Even though a "hunchbacked person" is not necessarily bad, society tends to look at them negatively as "deviant" from the norm.

Deviation How far a score is from the mean.

Deviation IQ A measure of intelligence-test performance based on an individual's standing relative to his own age-mates (e.g., an IQ of 100 is average and iqs of 70 and 130 correspond to percentile ranks of 2 and 98 respectively).

Dextrophobia Fear of objects at the right side of the body.

Diabetes A condition caused by a deficiency of insulin.

Diabetes mellitus A chronic form of diabetes involving an insulin deficiency and characterized by an excess of sugar in the blood and urine, and by hunger, thirst, and gradual loss of weight.

Diabetophobia Fear of diabetes.

Diagnosis The process of distinguishing among disorders.

Diagnostic and Statistical Manual of Mental Disorders (DSM) A reference book used by psychologists and psychiatrists to diagnose psychological disorders.

Diagnostic tests Tests of specific skills used to identify students1 needs and to guide instruction.

Diagnosticity In heuristics, the selection of evidence in order to facilitate choice between hypotheses.

Dialect The form or variety of a spoken language peculiar to a region, community, social group, occupational group, and so on.

Dialectical reasoning A process of going back and forth between opposing points of view in order to come up with a satisfactory solution to a problem.

Diary study A naturalistic complement to a think-aloud protocol in which subjects are asked to make their own record of their behaviour and feelings during a psychological experiment by keeping a log in a diary.

Diathesis-stress conception The belief that many organic and mental disorders arise from an interaction between a diathesis (a predisposition toward an illness) and some form of precipitating environmental stress.

Diathesis-stress hypothesis A hypothesis about the cause of certain disorders, such as schizophrenia, that suggests that genetic factors predispose an individual to a certain disorder, but that environmental stress factors must impinge in order for the potential risk to manifest itself.

Diathesis-stress model A causal account of mental disorders based on the idea that mental disorders develop when a person possesses a predisposition for a disorder, acquired through genetics and early learning experiences, and faces stressors that exceed his or her abilities to cope with them.

Dichotic listening A task that requires a person to attend to one of two different messages being presented simultaneously, one to each ear, through headphones.

Dichotic presentation An experimental procedure in which the participant hears two simultaneous messages, one presented to each ear. Typically, one of these is to be attended to and the other ignored.

Dichromat A person who is sensitive to only two of the three wavelengths of light.

Didactic learning An educational method in which a teacher lectures to students.

Didaskaleinophobia Fear of going to school.

Difference threshold The smallest physical difference between two stimuli that can still be recognized as a difference; operationally defined as the point at which the stimuli are recognized as different half of the time.

Differentiation A progressive change from the general to the particular and from the simpler to the more complex that characterizes embryological development. According to some theorists, the same pattern holds for the development of behaviour after birth.

Diffusion of responsibility An explanation of the failure of bystander intervention stating that when several bystanders are present, no one person assumes responsibility for helping.

Dikephobia Fear of justice.

Diminishing returns principle Applied to the perceived value of money, the principle states that the increase in the subjective value produced by every additional dollar decreases the more dollars the person has already. The same principle applies to the subjective value of other gains and losses. It also applies to the psychological magnitude of sensory qualities, as in the case of Weber's law.

Dinophobia Fear of dizziness or whirlpools.

Diplophobia Fear of double vision.

Dipsophobia Fear of drinking.

Direct aggression An attack placed upon what the individual believes is the source of their frustration. As opposed to Displaced Aggression.

Direct dyslexia A language disorder caused by brain damage in which people can read words aloud without understanding them.

Direct instruction Approach to teaching in which lessons are goal-oriented and structured by the teacher.

Direct perception An approach to vision, sometimes called ecological perception in which features of the stimulus are thought to correspond directly to features of the world, without the need for inference or high level knowledge of the world.

Direct relationship High values on one variable tend to occur with high values on the other variable.

Directed thinking Thinking that is aimed at the solution of a problem.

Directional A hypothesis which predicts that one group will have a higher or lower result than another group.

Directional selectivity A property of many visual neurones in which the neurone responds better to some directions of stimulus motion than to others.

Disability The inability to do something specific such as walk or hear.

Disassociation The functioning of consciousness at different levels without the awareness of relevant information on other levels.

Disavowal A splitting of an ego confronted by some distressing demand from the outer world. Contrast this with repression, where the demand comes from within.

Discipline Methods used to prevent behaviour problems from occurring or to respond to behaviour problems so as to reduce their occurrence in the future.

Disclosure reciprocity The tendency for one person's intimacy of self-disclosure to match that of a conversational partner.

Discontinuous theory of development Theory based on the belief that human development occurs through a series of distinct stages.

Discounting principle A principle of attribution theory stating that whenever there are several possible causal explanations for a particular event, people tend to be much less likely to attribute the effect to any particular cause.

Discovery learning Teaching methods in which students are encouraged to discover principles for themselves.

Discriminant validity The extent to which the test does not measure anything that is not in its intended concept—for example, some argue that an IQ test measures achievement, not potential, thus it measures the wrong concept.

Discrimination In classical conditioning, the appearance of a CR when one stimulus is presented (the CS+) but not another (the CS-). In operant conditioning, responding only when a specific discriminative stimulus is present but not when similar stimuli are present. In social psychology, the differential treatment of people based on their membership in a particular group.

Discriminative stimuli In instrumental conditioning, the external stimuli that signal a particular relationship between the instrumental response and the reinforcer. For example, a green light is a positive discriminative stimulus when it signals to a pigeon that it will get food if it hops on a treadle; the reverse is true of a red light, or the negative discriminative stimulus, which indicates that this action will not lead to a food reward.

Disease model of addiction The idea that addiction is a disease that has to be medically treated.

Disengagement theory views aging as a process of mutual withdrawal in which older adults voluntarily slow down by retiring, as expected by society.

Dishabiliophobia Fear of undressing in front of someone.

Disinhibition An increase of some reaction tendency by the removal of some inhibiting influence upon it (e.g., the increased strength of a frog's spinal reflexes after decapitation).

Disorganized An offender classification used in many criminal profiling systems. Such an offender tends to carry out spontaneous and impulsive acts in a manner that is sloppy and at high risk of leaving evidence. Such offenders tend to know the victims (at least by sight), stick to their own geographical locations, and use a blitz like assault with a weapon of convenience. Sexual acts often occur postmortem. SEE 'organized'.

Disorganized type A subtype of schizophrenia characterized by disor-ganized behaviour, disorganized speech, and emotional flatness or inappro-priateness.

Disorganized type of schizophrenia A subtype of schizophrenia in which the predominant symptoms are extreme incoherence of thought and marked ina-ppropriateness of behaviour and affect.

Disorientation Orientation is a function of the mind involving awareness of three dimensions: 1. time,

2. Place and

3. Person. Problems with orientation lead to disorientation, and can be due to various conditions, from delirium to intoxication. Typically, disorientation is first in time, then in place and finally in person.

Displaced aggression An attack against a person/object that is not the original source of frustration. This act will occur when the source of the frustration is either unavailable or is likely to retaliate.

Displacement The dream's tricky transfer of high-impact emotionality onto unimportant material and an emotional cooling to hot material. This and condensation are the most important doers of the distortive dreamwork that censors the true wishes underneath the dream. Displacement also refers to the tendency of libido to invest itself in objects other than the original object of its aim. It is therefore a key factor in personality development.

Display A term used by ethologists to describe genetically preprogrammed responses that serve as stimuli for the reaction of others of the same species, and thus serve as the basis of a communication system (e.g., mating rituals).

Display rule A culturally determined rule that prescribes the expression of emotions in particular situations.

Dispositional attribute An attribute explained or interpreted as being caused by internal influences.

Dispositional factors Individual personality characteristics that affect a person's behaviour.

Dispositional quality Any underlying attribute that characterizes a given individual and makes him more disposed than others to engage in a particular behaviour (e.g., the presence or absence of some ability or some personality trait).

Dispositional variables The organismic variables, or inner determinants of behaviour, that occur within human and nonhuman animals.

Dissociation The Diagnostic and Statistical Manual of Mental Disorders considers symptoms such as depersonalization, derealization, and psychogenic amnesia as core features of dissociation. However, in the normal population mild dissociative experiences are highly prevalent, with 80% to 90% of the respondents indicating that they experience dissociative experiences at least some of the time.

Dissociative amnesia A disorder characterized by an inability to remember extensive, important personal information, usually about something traumatic or painful.

Dissociative disorders A class of disorders in which anxiety is reduced by a sudden disruption in consciousness, which in turn produces changes in one's sense of identity.

Dissociative fugue A state in which the person wanders away from home, and then, days or even months later, suddenly realizes that he is in a strange place and does not know how he got there; this pattern is often understood as a means of coping with (and escaping from) extraordinarily painful events.

Dissociative identity disorder (DID) A dissociative mental disorder in which two or more distinct personalities exist within the same individual; formerly known as multiple personality disorder.

Dissociative mechanism Personality or mind splitting into two or more parts and functioning rather independently.

Dissonance Theory in the study of attitude functions, the suggestion that people are motivated to attend to and prefer information that is consistent with their current attitudes rather than information that is inconsistent (i.e. Dissonant).

Dissonance theory In the study of attitude functions, the suggestion that people are motivated to attend to and prefer information that is consistent with their current attitudes rather than information that is inconsistent (i.e. dissonant).

Distal stimulus An object or event outside (e.g., a tree) as contrasted to the proximal stimulus (e.g., the retinal image of the tree), which is the pattern of physical energies that originates from the distal stimulus and impinges on a sense organ.

Distance education Taking courses through alternative learning formats, such as intensive study classes conducted one weekend per month, telecourses provided over the television, or virtual classrooms set up on the Internet.

Distinctive feature A physical characteristic of an object that helps distinguish it from other objects.

Distinctiveness The phenomenon in memory that states we are better able to remember information if it is distinctive or different from other information.

Distortion When a subject does not respond honestly to questions.

Distraction conflict theory A theory holding that the presence of others will produce social facilitation effects only when those others distract from the task and create attentional conflict.

Distractors Incorrect responses offered as alternative answers to a multiple-choice question.

Distraught witness The distraught witness bears a great deal of emotional distress. This distress is the direct result of either witnessing the crime or indirectly from their relationship with the victim.

Distress calls The innate signals through which a human or animal infant indicates its need of aid.

Distressor A negative event, such as a death or loss of a job, that is stressful.

Distributed practice The practice of learning material in short sessions over a long period. It is also called the *spacing effect.*

Distributed representations A model of cognitive organization, especially semantic memory, in which each concept is represented, not by a designated node or group of nodes, but by a widespread pattern of activation across the entire network. *See also* connectionist models, local representations, network model, node.

Divergent thinking A style of thinking in which people's thoughts go off in different directions as they try to generate many different solutions to a problem.

Divorce The legal and formal dissolution of a marriage.

DNA Deoxyribonucleic acid, the structure of chromosomes. DNA structure resembles that of a twisted ladder. Strands of sugar and phosphates are connected by rungs made from adenine and thymine and guanine and cytosine.

Docility Among humans, the genetical trait of non-reciprocal altruism and acceptance.

Doctrine of specific nerve energies The law formulated by Johannes mÿller which holds that differences in sensory quality are not caused by differences in the stimuli themselves but by the different nervous structures that these stimuli excite. Thus, stimulating the retina will produce sensations of light, whether the retina is stimulated by a beam of light or pressure to the eyeball.

Dodo bird verdict An expression often used to summarize the comparison of the effectiveness of different forms of psychotherapy. According to the dodo bird in *Alice in Wonderland,* "Everyone has won and all must have prizes." Regarding psychotherapy, this statement is unders-tood to mean that all the major forms of psychotherapy are equally effective.

Domatophobia Fear of houses or being in a house.(Eicophobia, Oikophobia).

Dominance hierarchies A social order developed by animals that live in groups by which certain individuals gain status and exert power over others.

Dominant allele The form of the gene that controls the expression of a trait. When a gene pair contains two dominant alleles or when it contains both a dominant and recessive allele, the trait regulated by the dominant gene will be expressed.

Door-in-the-face technique A method for achieving compliance in which a certain request is preceded by a much larger one. The refusal of the first request, and the apparent concession on the part of the requester, makes people more likely to agree to the second demand, feeling that they should now make a concession of their own.

Dopamine Dopamine is a chemical naturally produced in the body. In the brain, dopamine functions as a neurotransmitter, activating dopamine receptors. Dopamine is also a neurohormone released by the hypothalamus. Its main function as a hormone is to inhibit the release of prolactin from the anterior lobe of the pituitary.

Dopamine hypothesis of schizophrenia Asserts that schizophrenics are oversensitive to the neurotransmitter dopamine. Evidence for this view comes from the fact that the classical antipsychotics, which alleviate positive schizophrenic symptoms, block dopamine transmission.

Dopamine-serotonin interaction hypothesis Asserts that schizophrenics are oversensitive to both dopamine and serotonin. Evidence for this view comes from the fact that atypical antipsychotics, which relieve both positive and negative symptoms, block receptors for both dopamine and serotonin.

Doraphobia Fear of fur or skins of animals.

Double-bind The conflict caused for a child when he or she is given inconsistent messages or cues from a parent.

Double blind study Double-blind describes an especially stringent way of conducting an experiment, usually on human subjects, in attempt to eliminate subjective bias on the part of both experimental subjects and the experimenters.

Double opponent cell A type of visual cell found, for example, in the lateral geniculate nucleus, that has a complex receptive field showing inhibition both between different wavelengths and different positions.

Double-blind technique A technique for evaluating drug effects independent of the effects produced by the expectations of research participants (placebo effects) and of physicians. This is done by assigning patients to a drug group or a placebo group with both patients and staff members in ignorance of who is assigned to which group.

Down syndrome A genetic disorder caused by a chromosomal aberration resulting in an extra 21st chromosome. People with Down syndrome are generally short with broad skulls and round faces, and suffer impairments in physical, psychomotor, and cognitive development.

Downward social comparison The defensive tendency to compare ourselves with others who are worse off than we are.

Doxophobia Fear of expressing opinions or of receiving praise.

Dramaturgical approach An approach used in the explanation of behaviour in social situations, where social life is seen as a metaphor for stage performance, with individuals selecting the roles and scripts to be played.

Drapetomania A fictitious mental illness believed to cause slaves to run away from their masters and obsessively seek freedom. An example of the misuse of the medical model of psychopathology.

Dream An invariably egotistical internal state made external. A brief nightly psychosis. A regression prompted by an infantile wish seeking fulfillment. "The royal road to the unconscious" because dream symbols (discovered by Scherner) can be interpreted in a way that enlightens the therapist about what's going on in the patient's unconscious. Dreams are compromises in that they express a conflict about a wish: pulling together material left over from the previous day, they fulfill the wish (or several wishes) symbolically but keep that unconscious and the dreamer asleep through the use of a repressive censor who rewrites the latent emotional material into manifest imagery difficult to understand. Dreams are invariably the products of conflict and can be prompted by either ego or id.

Note: Freud's work on dreams is often translated in a way that gives the impression of a fixed system of symbols (e.g., tunnels mean vaginas and towers mean penises).

Dream analysis The psychoanalytic interpretation of dreams used to gain insight into a person's unconscious mo-tives or conflicts.

Dream day The day before the dream, and from which it gathers its material.

Dream series A serious of similar dreams; it is to be interpreted as a whole.

Dreamwork The mental activity that translates the latent wish-seeking unconscious material into the manifest imagery that disguises it. Dreamwork is NOT the work one does on a dream. Includes condensation, displacement of affect, identification, composition, inversion, and secondary elaboration. (Since Freud's time the dreamwork seems to be falling down on the job. Some of these cigar-in-mouth and piston-in-shaft and tearing-your-boss's-head-off dreams don't seem to be very well disguised.).

Drill and practice Applications of microcomputers that provide students with practice of skills and knowledge.

Drive A condition, often caused by physiological changes or homeostatic disequilibrium, that energizes an orga-nism's behaviour.

Drive derivative Associations, symptoms, anxiety, and other kinds of material derived from a repressed drive.

Drive reduction hypothesis The hypothesis that a drive (resulting from physiological need or deprivation) produces an unpleasant state that causes an organism to engage in motivated behaviours. Reduction of drive is assumed to be reinforcing.

Drive reduction theories of motivation Ideas that suggest people act in order to reduce needs and maintain a constant physiological state.

Drive theories of aggression The theory suggesting that aggression is created by external conditions that arouse the motive to harm others. See "frustration aggression hypothesis".

Drive By drive Freud meant the bodily demands upon mental life. Freud believed in good materialist fashion that mind arose from drive, ego from id. A drive has a source (bodily needs that arise from the erogenous zones), an impetus, an internal aim (temporary removal of the bodily need), an external aim (the steps taken to reach the final goal of the internal aim), and an object. Drives give rise to the libido-energy that drives all psychological activity. We never experience the drive itself, just its representation or idea in the mind.

A drives might undergo the following "vicissitudes": reversal into its opposite (via either changing from active to passive or reversing its content—love into hate); turning around upon the subject; repression; sublimation; and anxiety. The operation of the first two vicissitudes depends on the narcissistic organization of the ego. All these vicissitudes amount to the drive subjugated to the three great polarities that regulate psychic life: activity/passivity (biological), ego/external world (real), pleasure/pain (economic).

Drive-reduction theory A theory that claims that all built-in rewards are at bottom reductions of some noxious bodily state. The theory has difficulty in explaining motives in which one seeks stimulation, such as sex and curiosity.

Dromophobia Fear of crossing streets.

Drug therapy Treatment that involves the use of medications. It is also called *pharmacotherapy*.

Drug tolerance The decrease in responsiveness to a drug developed after repeated use of a drug. Addicts must use increasingly larger doses to obtain the same effect that was produced previously.

DSM-III The diagnostic manual of the American Psychiatric Assocaition adopted in 1980. A major distinction between it and its predecessor is that it categorizes mental disorders by their descriptive characteristics rather than by theories about their underlying cause. Thus, a number of disorders that were formerly grouped together under the general heading "neurosis" (e.g., phobias, conversion disorders) are now classified under separate headings.

DSM-III-R The diagnostic manual of the American Psychiatric Association adopted in 1987, a relatively minor revision of its predecessor, DSM-III.

DSM-IV The current diagnostic manual of the American Psychiatric Association (adopted in 1994), a substantial revision of its predecessor, DSM-III-R.

DSM-IV-TR The current diagnostic and statistical manual of the American Psychiatric Association that classifies, defines, and describes mental disorders.

Dual attitudes Differing implicit (automatic) and explicit (consciously controlled) attitudes toward the same object. Verbalized explicit attitudes may change with education and persuasion; implicit attitudes change slowly, with practice that forms new habits.

Dual code theory of memory Theory suggesting that information coded both visually and verbally is remembered better than information coded in only one of those two ways.

Dual-center theory A hypothesis about the hypothalamic control of eating. One center (in the lateral hypothalamus) was hypothesized as the "on" center, the initiator of eating; another center (in the ventromedial region) was hypothesized as the "off" center, the terminator of eating. Current evidence indicates, however, that these brain regions, while crucial for eating, are only a part of the circuits controlling eating.

Dualism The concept, discussed by various philosophers since descartes, that the mind and the brain (i.e. Matter) are dual entities which have separate functions, with the mind exerting an influence on the physical workings of the brain that in turn control the muscles of the body.

Duplex theory of vision The theory that rods and cones handle different aspects of vision. The rods are the receptors for night vision; they operate at low light intensities and lead to achromatic (colourless) sensations. The cones are used in day vision; they respond at higher illumination levels and are responsible for sensations of colour.

Dutchphobia Fear of the Dutch.

Dyad An encounter between two individuals.

Dyslexia A condition in which the high-speed language processing area of the brain fails to activate, causing difficulty in reading and deciphering particular sounds.

Dysmorphophobia Fear of deformity.

Dysthymic disorder A disorder involving depressed mood on a majority of days for at least two years.

Dystychiphobia Fear of accidents.

E

Eardrum The taut membrane that transmits vibrations caused by sound waves across the middle ear to the inner ear.

Early adult transition A stage in the novice phase of early adulthood, ranging from ages 17 to 22.

Early adulthood ages 17 to 45.

Early childhood ages 2 to 6.

Early intervention programs Compensatory preschool programs that target very young children at the greatest risk of school failure.

Early phase of labor The phase of labor that consists of mild, minute-long contractions that occur every 15 minutes.

Eating disorders Disorders characterized by problematic eating patterns, extreme concerns about body weight, and inappropriate behaviours aimed at controlling body weight.

Ecclesiophobia Fear of church.

Echoic memory A form of sensory memory for sounds that have just been perceived.

Echoic memory Sensory memory that allows auditory information to be stored for brief durations.

Eclampsia An attack of convulsions; specifically, a disorder that may occur late in pregnancy, characterized by convulsions, edema, and elevated blood pressure.

Eclectic approach A form of therapy in which the therapist uses whatever method he or she feels will work best for a particular client at a particular time.

Ecological approach In environmental psychology, the study of behaviour settings in order to predict their effect on particular individuals.

Ecomania A pathological attitude direct towards one's family. It is characterized by domineering behaviour. This attitude has implications in familial abuse.

Ecophobia Fear of home.

Ectoderm The outer layer of cells of an embryo from which the nervous system, skin, hair, teeth, and so on are developed.

Ectopic pregnancy A pregnancy with the fertilized ovum developing outside the uterus, as in a fallopian tube.

Ectotherms Organisms that control their body temperature by using mechanisms that are mostly external (such as choosing a sunny or shady environment). Previously called *cold blooded*.

Educational Psychology The study of learning and teaching.

Effect size An effect size is the strength or magnitude of the difference between two sets of data or, in outcome studies, between two time points for the same population. (The degree to which the null hypothesis is false).

Effectors Organs of action; in humans, muscles and glands.

Effects of content In the solving of syllogisms, the effects of prior beliefs and knowledge relating to the basic information presented, which may affect the process of syllogistic reasoning.

Efferent nerves Bundles of axons that carry information from the central nervous system to muscles and sense organs.

Effort after meaning The amount of effort utilized by witnesses to describe a recent specific event or a particular person's behaviour.

Effortful processing Practicing or rehearsing information through either shallow or deep processing.

Ego (Ich) The "I," a rational, organized agency that distills gradually out of a passionate id that rubs up against reality. Emerging from an undifferentiated mass of sensations (chiefly those emanating from the surface of the body), formed by identifications and abandoned id cathexes, and strengthened by speech, which links auditory and visual memory traces with the conscious life, the ego strives to harmonize inner and outer, drives (which it keeps at bay mainly via repression, sublimation and anticathexes), inhibition, and reality. Unlike Jung's conception of ego, it is not entirely conscious, though it is the bearer of consciousness. Topographically, it extends down into the id.

Note: the Latinized term ego is unnecessary and mechanical; Freud wrote, and meant, "I.".

Ego cathexes Investments of libido into ego activities like ideas that the ego knows represent objects. You think about what you want instead of blindly leaping for it because of the energy investment remaining with your thoughts.

Ego defence mechanisms Mental strategies (conscious or unconscious) used by the ego to defend itself against conflicts experienced in the normal course of life.

Ego development A collective term for the various stages of life during which a human being acquires and masters those functions necessary to deal with the world at large. The American psychoanalyst, Erik Erikson, has described ego development in terms of eight distinct, psychosocial stages which a human moves through during a lifetime life.

Ego ideal The sector of the superego that consists of internalized standards of what is "good." Shoulds and oughts for Freud represent the higher human nature, whereas for most of us they represent the highest accomplishment of Sunday school.

Ego instincts An early concept denoting the ego's self-preservative drives (as opposed to the other-directed sexual drives). The concept came about from the notion of conflicts between ego and sexuality. The duality of Eros and the death drive replaced this old duality.

Ego integrity Maintaining one's sense of wholeness.

Ego psychology An approach to psychology that, in addition to the neo-Freudian concern with cultural and interpersonal factors, stresses the healthy aspects of the self as it tries to cope with reality.

Egocentric Egocentrism is the practice of regarding oneself and one's own opinions or interests as most important. The term is derived from the Greek "I".

Egocentric orientation In environmental psychology, the child's understanding of its environment only as it relates to self, and which is characteristic of the sensorimotor stage of development.

Egocentric viewing Everything in relation to oneself; self-centred.

Egocentrism In cognitive development, the inability of a young child at the preoperational stage to take the perspective of another person.

Egoism A motive (supposedly underlying all behaviour) to increase one's own welfare. The opposite of altruism, which aims to increase another's welfare.

Egoistic Motivated by the desire to increase one's own welfare.

Egoistic helping A form of helping in which the ultimate goal of the helper is to increase his or her own welfare.

Eicophobia Fear of home surroundings. (Domatophobia, Oikophobia).

Eidetic memory A relatively rare kind of memory characterized by relatively long-lasting and detailed images of scenes that can be scanned as if they were physically present.

Eisoptrophobia Fear of mirrors or of seeing oneself in a mirror.

Elaboration A type of deep processing in which information being learned is associated with other meaningful material.

Elaboration likelihood model of persuasion (ELM) A theory which tries to account for the contradictory findings in the acceptance of persuasive communications by asserting that acceptance of a message can be achieved by two routes, the central and the peripheral, either of which depends on the extent to which the person thinks about and evaluates the message.

Elaborative rehearsal Rehearsal in which material is actively reorganized and elaborated while in working memory. In contrast to maintenance rehearsal, this confers considerable benefit.

Elderly abuse The neglect and/or physical and emotional abuse of dependent elderly persons.

Electra complex The feminine equivalent of the male Oedipal Complex. A chief difference is that the girl isn't required to shift her identification from her mother to her father. Also, because of penis envy, she wants to have a baby with her father; this desire later gets acted out by her choice of a fatherly man to have a baby with. The term "Electra Complex" may have been proposed by Jung.

Electric stimulation of the brain An invasive method of studying the brain, in which an implanted electrode activates a particular brain structure.

Electrocardiogram (EKG) The measurement and graphical presentation of the electrical activity of the heart, recorded by means of electrodes attached to the skin.

Electroconvulsive shock treatment (ECT) A somatic treatment, mostly used for cases of severe depression, in which a brief electric current is passed through the brain to produce a convulsive seizure.

Electroencephalogram (EEG) The process of recording the brain's electrical activity by attaching electrodes to the scalp and analyzing wave patterns.

Electromyograph (EMG) An instrument that records muscle activity.

Electro-oculogram (EOG) The measurement and graphical presentation of the electrical activity caused by movements of the eye, recorded by means of electrodes attached to the skin adjacent to the eye.

Electrophobia Fear of electricity.

Eleutherophobia Fear of freedom.

Elevation A monocular cue of depth perception; objects nearer the horizon are seen as farther from the viewer.

Elimination by aspects The process of eliminating alternatives in a decision based on whether they do or do not possess aspects or attributes the decision maker has deemed necessary or desirable.

Ellis, Albert Albert Ellis | Ellis, Albert(born September 27, 1913, Pittsburgh, Pennsylvania) is a psychologist whose Rational Emotive Behaviour Therapy (REBT), is the foundation of all cognitive and cognitive behaviour therapies. REBT is a comprehensive theory of personality and psychotherapy which holds that one's personal beliefs, evaluations, and personal philosophy control one's feelings.

Elurophobia Fear of cats. (Ailurophobia).

Embarrassment An unpleasant emotion experienced when we believe that we cannot perform coherently in a social situation.

Embedded sentence A sentence structure in which one full sentence is included in the midst of another, as in "The girl who ate the hamburger hit the ball." Here the sentence "[she] ate the hamburger" interrupts the flow of the main sentence, "The girl . . . Hit the ball.".

Embryo An animal in the earliest stages of its development in the uterus or the egg, specifically, in humans, from conception to about the eighth week.

Embryonic stage The period that begins two weeks after conception and ends two months after conception.

Emergency reaction Intense sympathetic arousal that mobilizes an organism for a crisis.

Emergent literacy Knowledge and skills relating to reading that children usually develop from experience with books and other print media before the beginning of formal reading instruction in school.

Emetophobia Fear of vomiting.

Emotion A complex pattern of changes, including physiological arousal, feelings, cognitive processes, and behavioural reactions, made in response to a situation perceived to be personally significant.

Emotion focused coping Cognitive and behavioural efforts to reduce the distress produced by a stressful situation.

Emotion work The process of acting out of an emotion that is not really felt.

Emotional aggression Inflicting harm for its own sake.

Emotional and behavioural disorders Category of exceptionality characterized by problems with learning, interpersonal relationships, and control of feelings and behaviour.

Emotional intelligence(EQ) Type of intelligence defined as the abilities to perceive, appraise, and express emotions accurately and appropriately, to use emotions to facilitate thinking, to understand and analyze emotions, to use emotional knowledge effectively, and to regulate one's emotions to promote both emotional and intellectual growth.

Emotional stability The tendency to be relaxed and at peace with oneself.

Empathic concern A feeling of sympathy and concern for the sufferings of another coupled with the desire to relieve this suffering.

Empathy The projection of one's own personality into the personality of another in order to understand the person better; ability to share in another's emotions, thoughts, or feelings.

Empathy altruism hypothesis The proposition that empathic concern for a person in need produces an altruistic motive for helping.

Empirical evidence Recorded observations which together form the basis of psychology as a science and which can be inspected and publicly questioned.

Empirically validated treatments Treatments that are shown by research to be more effective for a particular problem than a placebo or no treatment.

Empiricism A school of thought that holds that all knowledge comes by way of empirical experience, that is, through the senses.

Empty nest A popular term for that time in an adult's life when they have to come to terms with their sense of loss and loneliness when their children grow up and leave home. Encoding the extraction of information from a stimulus in order to form a memory trace.

Empty nest syndrome A form of mental depression said to be caused in parents by the loss felt when their children grow up and leave home.

Enactment Learning process in which individuals physically carry out tasks.

Encephalization Increases in brain size.

Encoding The process by which a mental representation is formed in memory.

Encoding specificity principle A proposition that retrieval is a function of the overlap between a memory trace and the information provided at retrieval, which led to the view that recall and recognition are different aspects of the same process.

Encoding specificity The principle that subsequent retrieval of information is enhanced if cues received at the time of recall are consistent with those present at the time of encoding.

Endocrine gland A gland that secretes a hormone.

Endocrine system The system of ductless glands whose secretions are released directly into the bloodstream and affect organs elsewhere in the body (e.g., adrenal gland).

Endoderm The inner layer of cells of the embryo, from which the lining of the digestive tract, other internal organs, and certain glands are formed.

Endogenous biological rhythms Biological cycles that originate from inside the body rather than depend on cues from the environment.

Endometrium The inner lining of the uterus.

Endopsychic conflict The opposition of a cathexis by an anti-cathexis (or: being torn between a desire to satisfy a wish and a fear of doing so). Two types: id-ego conflicts and superego-ego conflicts.

Endorphin A drug produced within the brain itself whose effects and chemical composition are similar to such pain-relieving opiates as morphine.

Endotherms Organisms that control their body temperature by using mechanisms that are mostly internal or physiological. Previously called *warm blooded*.

Enetophobia Fear of pins.

Engaged time Time students spend actually learning; same as time on-task.

English as a second language Designation for programs and classes to teach English to students who are not native speakers of English.

Engram The hypothesized chemical change in the brain resulting from the storing of memory information; also called memory traces.

Enochlophobia Fear of crowds.

Enosiophobia or enissophobia Fear of having committed an unpardonable sin or of criticism.

Enrichment activities Assignments or activities designed to broaden or deepen the knowledge of students who master classroom lessons quickly.

Enrichment programs Programs in which assignments or activities are designed to broaden or deepen the knowledge of students who master classroom lessons quickly.

Entering the adult world A stage in the novice phase of early adulthood, ranging from ages 22 to 28.

Entomophobia Fear of insects.

Entrapment The condition in which commitments to a failing course of action are increased to justify investments already made.

Envelope The overall shape of a waveform.

Environment, non shared Environment that is unique to the individual, such as the individual's peer group.

Environment, shared Environment that is shared by siblings reared in the same family, such as parents' education and socioeconomic status.

Environmental cognition In environmental psychology, the individual's perception of the real world at molar (i.e. Comprehensive) rather than molecular level.

Environmental determinism In environmental psychology, those everyday, external aspects of living, such as good or bad housing, which make the occurrence of a certain behaviour more likely.

Environmental grammar The terms of reference acquired by an individual in relation to the spaces and settings around them, from which they develop the appropriate expectancies and hence, the corresponding search strategies.

Environmental programming In the design of buildings, the systematic collection of data on user-needs in order to understand better their collective needs.

Environmental simulations In environmental programming, the use of certain communication tools, such as drawings and models, in order to elicit responses from potential users of buildings etc.

Environmental variables External influences on behaviour.

Enzymes Proteins that regulate the structure of bodily cells and the processes occurring within those cells.

Eosophobia Fear of dawn or daylight.

Ephebiphobia Fear of teenagers.

Epidemiology The study of how diseases spread, the effect that they have on the environment and how they are controlled.

Epinephrine (adrenaline) A neurotransmitter released into the bloodstream by the adrenal medulla as part of sympathetic activation (e.g., racing heart).

Episiotomy Incision of the perineum, often performed during childbirth to prevent injury to the vagina.

Episodic memories Long-term memories for autobiographical events and the contexts in which they occurred.

Episodic memory A part of long-term memory that stores images of our personal experiences.

Epistaxiophobia Fear of nosebleeds.

Epistemophobia Fear of knowledge.

Equal intervals Characteristic of a scale of measurement where the individual units possess the qualities of equal intervals. The difference between each unit of measurement is exactly the same.

Equal-status contact Contact on an equal basis. Just as a relationship between people of unequal status breeds attitudes consistent with their relationship, so do relationships between those of equal status. Thus, to reduce prejudice, interracial contact should be between persons equal in status.

Equilibration The process of restoring balance between present understanding and new experiences.

Equilibrium mode In the dyad, the balancing of the level of intimacy through the gaze pattern, which is adjusted according to the distance the two people are away from each other.

Equinophobia Fear of horses.

Equipotentiality The claim (contradicted by much evidence) that organisms can learn to associate any response with any reward or to associate any pair of stimuli.

Equity A condition in which the outcomes people receive from a relationship are proportional to what they contribute to it. Note: equitable outcomes needn't always be equal outcomes.

Equity theory The theory that people are most satisfied with a relationship when the ratio between benefits and contributions is similar for both partners.

Eremophobia Fear of being oneself or of lonliness.

Ereuthrophobia Fear of blushing.

Ergasiophobia .1 Fear of work or functioning. .2 Surgeon's fear of operating.

Ergonomics The study of the relationship between the individual worker and the demands of their job and their working environment, paying particular attention to the efficiency with which their work is performed.

Ergophobia Fear of work.

Erogenous (or Erotogenic) zones Those areas of the body most liable to sexual stimulation—namely, the mouth, anus, and penis, although Freud regarded the entire body as an erogenous zone.

Eros One of the two basic sources of all the drives. Eros, whose name comes from the Greek god of love, is the principle of life; it binds together and is most clearly seen in love. Its drives tend to be more plastic and displaceable than those of its opponent, Thanatos, the death drive. Freud saw psychic life as an interplay of these two ever-interpenetrating forces, Life and Death.

Erotogenicity The activity of a bodily area that conveys sexually exciting stimuli to the mind.

Erotomania A pathological exaggerated sexual interest. In males this condition is satyriasis; in females, nymphomania.

Erotophobia Fear of sexual love or sexual questions.

Error The amount of other variables (aside from what you are measuring) that can impact the observed score.

Error level The level of accepted error within a given set of data. The greater the error level, the wider the confidence interval.

Erythrophobia or erytophobia or ereuthophobia .1 Fear of redlights.

2 Blushing.

3. Red.

Escalation effect The tendency for investors to remain committed to a losing course of action.

Escape behaviour Behaviour that terminates an ongoing event.

Escape conditioning Operant conditioning based on the idea that a behaviour is more likely to be repeated if it results in the cessation of a negative event.

Escape learning Instrumental learning in which reinforcement consists of the reduction or cessation of an aversive stimulus (e.g., electric shock).

Escape response An operant response acquired through negative reinforcement that terminates an aversive stimulus.

Escapism Escapism is mental diversion by means of entertainment or recreation, as an "escape" from the unpleasant aspects of daily reality.

Estimate An idea about a characteristic of a population based on sample data (e.g., the sample mean IQ was 102 so we estimate that the population mean IQ is also 102).

Estrogen Any of several female sex hormones that cause estrus; estrogen helps to stimulate enlargement of the reproductive organs and relaxation of associated ligaments, stimulate development of the uterine lining and mammary glands, and prevent contractions of the uterus.

Estrous cycle The ovulatory cycle in mammals other than primates; the sequence of physical and hormonal changes that accompany the ripening and disintegration of ova.

Estrus In mammals, the period in the cycle when the female is sexually receptive (in heat).

Eta A correlational technique used primarily for non-linear relationships. (Example, income and age are positively correlated until older age at which point the correlation reverses itself to some extent.

Ethical dilemma A situation in which a researcher must make a difficult moral decision.

Ethics A system of moral values.

Ethnic group A group within a larger society that sees itself as having a common history, social and cultural heritage, and traditions, often based on race, religion, language, or national identity.

Ethnic identity An individual's sense of personal identification with a particular ethnic group.

Ethnicity A history, culture, and sense of identity shared by a group of people.

Ethnocentrism A belief in the superiority of one's own ethnic and cultural group, and a corresponding disdain for all other groups.

Ethnographic research An intensive (case) study of a group.

Ethology A branch of zoology concerned with the behaviour of animals under natural conditions.

Etiology Causal relationships of diseases; theories regarding how the specific disease or disorder began.

Eunuch A castrated male.

Euphobia Fear of hearing good news.

Eurotophobia Fear of female genitalia.

Eustressor A positive event, such as marriage or vacations, that is stressful.

Euthanasia Act or practice of causing death painlessly in order to end suffering advocated by some as a way to deal with persons dying of incurable, painful diseases.

Evaluation Decision making about student performance and about appropriate teaching strategies.

Evaluation apprehension Concern for how others are evaluating us.

Evaluation apprehension theory A theory holding that the presence of others will produce social facilitation effects only when those others are seen as potential evaluators.

Event related potentials (ERP) An instrument used to measure and record brainwaves during specific cognitive events.

Events of instruction A model of instruction developed by Gagné that matches instructional strategies with the cognitive processes involved in learning.

Everyday reasoning A form of reasoning that is linked with practical action and is based on people's ability to develop convincing arguments about everyday issues, and which is based on situation modelling.

Evolution A change in the frequency of genes in a population.

Evolutionary perspective The approach to psychology that stresses the importance of behavioural and mental adaptiveness, based on the assumption that mental capabilities evolved over millions of years to serve particular adaptive purposes.

Evolutionary psychology The branch of psychology that studies the ways in which an organism's evolutionary history contributes to the development of behavioural patterns and cognitive strategies related to reproduction and survival during its lifetime.

Evolutionary stable strategies (ESS) In sociobiology, a strategy for species preservation, in the form of an adaptive behaviour, which can resist invasion by other strategies.

Ex post facto (After the fact) research Research method in which the independent variable is administered prior to the study without the researcher's control and its effects are investigated afterward.

Exceptional learners Students who have abilities or problems so significant that the students require special education or other services to reach their potential.

Exchange relationship A hypothesized type of social relationship in which the relationship depends on reciprocity; if goods (or esteem or loyalty) are given by one of the partners in the relationship, then the other must respond in kind.

Excitation threshold The voltage difference between a neuron's interior and exterior that, if exceeded, causes the neuron to fire. This voltage is about -55 millivolts in mammals. If the voltage reaches this threshold (from a "resting" voltage of -70 millivolts), the neuron's membrane destabilizes, leading to an action potential.

Excitation transfer The process whereby arousal caused by one stimulus is added to arousal from a second stimulus and the combined arousal is attributed to the second stimulus.

Excitation transfer effects The transfer of autonomic arousal from one situation to another, as when strenuous exercise leads to an increased arousal when presented with aggression-arousing or erotic stimuli.

Excitatory inputs Information entering a neuron that signals it to fire.

Excitatory postsynaptic potential A positive change in voltage that occurs when a neurotransmitter binds to an excitatory receptor site.

Exemplar A memory of particular examples of objects or situations that are used as the basis of classifying objects or situations into concepts.

Exemplification Eliciting perceptions of integrity and moral worthiness.

Exhibitionism A compulsion to expose parts of the body, most often the sex organs, for the purpose of sexual excitement.

Existential anxiety Emotional troubles that arise from the denial of death.

Existential psychology The search for meaning in life through the study of death.

Existential therapy A humanistic therapy that emphasizes people's free will and tries to help them achieve a personal outlook that will give meaning to their lives.

Existentialism Philosophical approach that emphasizes individual freedom and responsibility in the face of limitations, including the ultimate limitation of death.

Expectancy The belief that a certain consequence will follow a certain action.

Expectancy effects Results that occur when a researcher or observer subtly -to participants the kind of behaviour he or she expects to find, thereby creating that expected reaction.

Expectancy theory A cognitive theory of work motivation that proposes that workers are motivated when they expect their efforts and job performance to result in desired outcomes.

Expectancy-value model In the explanation of social attitudes, a model devised as a uni-dimensional, or one-component, view of attitudes which provides a useful basis for understanding why different people hold different attitudes towards the same entity.

Expected value The process of adding the value of a win times the probability of a win to the value of a loss times the probability of a loss in order to make a decision.

Experience-sampling method An experimental method that assists researchers in describing the typical contents of consciousness; participants are asked to record what they are feeling and thinking whenever signaled to do so.

Experiential intelligence According to Sternberg, the ability to deal effectively with novel situations and to automatically solve problems that have been encountered previously.

Experiment A form of research that can demonstrate causal relationships because

1. the experimenter has control over the events that occur and.

2. Participants are randomly assigned to conditions.

Experimental ablation The removal or destruction of a portion of the brain of an experimental animal for the purpose of studying the functions of that region.

Experimental aesthetics In cognitive psychology, the examination in detail of the relationship between particular physical and environmental features and people's aesthetic response to them.

Experimental group A group of subjects in an experiment, the members of which are exposed to a particular value of the independent variable, which has been manipulated by the experimenter.

Experimental method Research method using random assignment of subjects and the manipulation of variables in order to determine cause and effect.

Experimental neuropsychology The branch of psychology that attempts to understand human brain functions by studying patients whose brains have been damaged through accident or disease.

Experimental realism The degree to which experimental procedures are involving to participants and lead them to behave naturally and spontaneously.

Experimental research Studies that seek clues to cause-effect relationships by manipulating one or more factors (independent variables) while controlling others (holding them constant).

Experimenter bias Expectations of an outcome may inadvertently influence participant or cause the experimenter to view data in a different way.

Experimenter expectancy effects The effects produced when an experimenter's expectations about the results of an experiment affect his or her behaviour toward a participant and thereby influence the participant's responses.

Expert power Power derived through advanced knowledge or experience in a particular subject.

Expert systems Computer problem-solving programs with a very narrow scope that only deal with problems in a limited domain of knowledge (e.g., the diagnosis of infectious diseases).

Explicit attitudes Conscious beliefs that can guide decisions and behaviour.

Explicit memory Memory retrieval in which there is awareness of remembering at the time of retrieval.

Explicit uses of memory Conscious efforts to recover information through memory processes.

Exposure therapies Therapies that aim to eliminate anxiety responses by having clients face real or imagined versions of feared stimuli.

Expressed emotion Expressions of criticism, hostility, and emotional overinvolvement by family members toward a person with schizophrenia.

Expressive aphasia A disorder in which the patient has difficulty with the production of speech. Expressive aphasia is caused by a cortical lesion that damages one's ability to organize the movements necessary for speech production into a unified sequence.

Expressive language The ability to use language to communicate.

External attribution An attribution that locates the cause of an event to factors external to the person, such as luck, or other people, or the situation.

External locus of control The tendency to believe that circumstances are not within one's control but rather are due to luck, fate, or other people.

External validity The degree to which there can be reasonable confidence that the results of a study would be obtained for other people and in other situations.

Externality hypothesis The hypothesis that some and perhaps all obese people are relatively unresponsive to their own internal hunger state but are much more susceptible to signals from without.

Extinction In classical conditioning, the gradual weakening and disappearance of a conditioned response when a conditioned stimulus is not followed by an unconditioned stimulus. In operant conditioning, it's the gradual disappearance of a response after it stops being reinforced.

Extinction burst The increase in levels of behaviour in the early stages of extinction.

Extralinguistic factors Factors outside of the language itself that influence (and usually ease) language comprehension. Examples include gestures and the scene that is being discussed.

Extramarital affair Having sexual intercourse with someone other than one's spouse.

Extraneous variable A variable unrelated to an experiment (such as room temperature or noise level) that may interfere with the results of the experiment.

Extrapyramidal system One of the two cerebral motor control systems; it is older in evolutionary terms, and it controls relatively gross movements of the head, limbs, and trunk.

Extraversion Jung's concept for a person whose mind is externally oriented. In modern psychology this has come to mean a person who is very active, sociable and dominant.

Extrinsic incentive A reward that is external to the activity, such as recognition or a good grade.

Extrinsic motivation Traditionally, extrinsic motivation has been used to motivate employees: Payments, rewards, control, or punishments. Within economies transitioning from assembly lines to service industries, the importance of intrinsic motivation rises.

Extroversion Personality style where the individual prefers outward and group activity as opposed to inward and individual activity.

Extroversion-introversion With neuroticism-normality, one of two orthogonal dimensions applied in the description of personality, as proposed by the psychologist hans eysenck, and used in the statistical analysis of behavioural data, where variations in behaviour are defined between these two opposite, end points.

Eye movement desensitization and reprocessing (EMDR) A type of exposure therapy in which clients move their eyes back and forth while recalling memories that are to be desensitized.

Eye tracker A device that measures the location of a person's gaze while he or she observes a visual display.

Eysenck personality inventory (EPI) A questionnaire devised by the psychologist hans eysenck to capture the dimensions of personality in a questionnaire format for use in psychiatric diagnosis, using the extroversion-introversion and neuroticism-normality dimensions.

F

Face validity The degree to which test items appear to be directly related to the attribute the researcher wishes to measure.

Facial electromyograph (EMG) An electronic instrument that records facial muscle activity associated with emotions and attitudes.

Facial-feedback hypothesis The idea that the brain uses feedback from facial muscles to recognize emotions that are being experienced.

Factor Hypothetical construct that explains the correlations among observed variables, such as items on a test.

Factor analysis A statistical method for studying the interrelations among various tests, the object of which is to discover what the tests have in common and whether these communalities can be ascribed to one or several factors that run through all or some of these tests.

Factor loading Correlation of an item with a factor.

Factorial ANOVA An Analysis of Variance used when there are two or more independent variables. When there are two, the ANOVA is called a Two-Way ANOVA, three independent variables would use a Three-Way ANOVA, etc.

Fallopian tube Either of two slender tubes that carry ova from the ovaries to the uterus.

False belief tasks A series of tasks used in the study of theory of mind which provide a test case for finding out whether children really understand the relationship between beliefs and behaviour.

False consensus The tendency of a person to perceive his or her own views as representative of a general consensus.

False consensus bias The tendency to exaggerate how common one's own characteristics and opinions are in the general population.

False consensus effect In the study of the attribution process, the tendency to use our own attitudes and behaviour as the basis for deciding consensus for that behaviour, by overestimating the number of people who share our beliefs and habits.

False uniqueness An aspect of the false consensus effect, where the subject underestimates the number of people who share their particular skills and abilities.

False uniqueness effect The tendency to underestimate the commonality of one's abilities and one's desirable or successful behaviours.

False-belief tasks A series of tasks used in the study of theory of mind which provide a test case for finding out whether children really understand the relationship between beliefs and behaviour.

Falsifiability The ability of a theory or hypothesis to be rejected.

Falsifiable theory A theory capable of generating hypotheses that show it to be false and which is an essential feature of scientific method.

Falsification In experiments with selection tasks, a level of insight into the logical structure of the task demonstrated by the choosing of cards that are falsifying rather than confirming instances.

Familicide When an individual kills his/her spouse and one or more of his/her children.

Family of origin The family one was born into or raised by.

Family resemblance structure Overlap of features among members of a category such that no members of the category have all of the features but all members have some of them.

Family romance The tendency to defensively glamorize one's family of origin in order to avoid hating them.

Family studies Studies in which researchers examine trait similarities among members of a family to figure out whether that trait might be inherited.

Family therapy A general term for a number of therapies that treat the family or couple, operating on the assumption that the key to family or marital distress is not necessarily in the pathology of any individual family member but is rather in the interrelationships within the family.

Fast mapping The process whereby the child learns new nouns and verbs by relating a novel object or action to a novel word via the formation of a new concept in as little as a single experience.

Fausse reconnaisance A false memory of having told the analyst something previously. Prompted by a fantasy of wanting to tell something but not following through, perhaps due to **resistance**.

Fear A rational reaction to an objectively identified external danger that may induce a person to flee or attack in self-defence.

Fear induced aggression Responses believed to be biologically programmed into us so that we act in an aggressive manner towards any form of forced confinement.

Feature detector The idea deriving from early neuro-physical studies of the visual cortex that an individual cell (specifically a simple cell) could signal the presence of an individual line or edge of a specific orientation at a specific position in the image.

Feature detectors Neurons in the retina or brain that respond to specific features of the stimulus, such as movement, orientation, and so on.

Feature net A model of pattern recognition in which there is a network of detectors, with feature detectors at the bottom.

Feature-tracking The process thought to be used by the long range motion system by which image motion is recovered by tracking a distinct feature of the image from one moment to the next.

Febriphobia or Fibriphobia or Fibriophobia Fear of fever.

Fechner's law The assertion that the strength of a sensation is proportional to the logarithm of physical stimulus intensity.

Fee for service An arrangement for health care in which people pay providers for health care services.

Feedback Information on the results of one1s efforts.

Feedback system A system in which some action produces a consequence that affects (feeds back on) the action. In negative feedback, the consequence stops or reverses the action (e.g., thermostat-controlled furnace). In positive feedback, the consequence strengthens the action (e.g., rocket that homes in on airplanes).

Feigned scarcity Implying that a product is in scarce supply, even when it is not, in order to increase demand for it.

Felinophobia Fear of cats. (Ailurophobia, Elurophobia, Galeophobia, Gatophobia).

Femininity Possession of expressive personality traits.

Foetal alcohol syndrome A condition affecting infants, characterized variously by mental retardation, heart defects, physical malformations, and so on and caused by excessive consumption of alcohol by the mother during pregnancy.

Foetal stage The third and final stage of prenatal development, which lasts for about 7 months, beginning with the appearance of bone tissue and ending with birth.

Fetish A condition in which arousal and/or sexual gratification is attained through inanimate objects (shoes, pantyhose) or non-sexual body parts (feet, hair). Is considered a problem when the object is needed in order to obtain arousal or gratification and the individual can not can not complete a sexual act without this object present.

Fetishism A pathological condition in which sexual arousal and gratification is induced by the handling of objects or nonsexual parts of the body.

Foetus The unborn young of an animal while still in the uterus or egg, especially in its later stages and specifically, in humans, from about the eighth week after conception until birth.

Fiedler's contingency theory In the study and assessment of leadership amongst social groups, a method which takes account of the leader's personal style appropriate to a given situation.

Field dependence Cognitive style in which patterns are perceived as whole.

Field experiment An experiment that is more naturalistic in content than that performed in a laboratory, in that it is in less danger of distorting the evidence by taking the phenomenon out of its true context.

Field independence Cognitive style in which separate parts of a pattern are perceived and analyzed.

Field research Research done in natural, real-life settings outside the laboratory.

Fight-or-flight response A sequence of internal activities triggered when an organism is faced with a threat; prepares the body for combat and struggle or for running away to safety; recent evidence suggests that the response is characteristic only of males.

Figure Object-like regions of the visual field that are distinguished from background.

Figure ground relationship Perceiving selected parts of a stimulus to stand out (figure) from other parts (background).

Figure-ground organization The segregation of the visual field into a part (the figure) that stands out against the rest (the ground).

File-drawer problem A tendency for disappointing or negative results not to be reported (and so merely dumped into a file drawer). This tendency can cause a bias in the pattern of evidence available.

Fimbria A fringe or border of hairs, fibres, and so on or a fringelike process, especially at the opening of an oviduct in mammals.

Final common path The single neural output upon which two groups of nerve fibres converge.

Fine motor skills The use of small bodily movements, such as drawing or writing.

Finite state machine A simple theoretical mechanism which shows how each word in a string of words can be related to the next.

Five-factor model A comprehensive descriptive personality system that maps out the relationships among common traits, theoretical concepts, and personality scales; informally called the Big Five.

Fixation A brief interval between saccadic eye movements during which the eye does not move; visual information is gathered during this time. In Freud's view, an unconscious obsession with an erogenous zone resulting from failure to resolve the crisis associated with the corresponding stage of psychosexual development.

Fixed interval schedule Reinforcement schedule in which desired behaviour is rewarded following a constant amount of time.

Fixed ratio (FR) schedule Reinforcement schedule in which desired behaviour is rewarded following a fixed number of behaviours.

Fixed system of reference In environmental psychology, the child's use of known locations - home, friends' houses, shops, school - to provide an understanding of large-scale place.

Fixed-action patterns Term used by ethologists to describe stereotyped, species-specific behaviours triggered by genetically preprogrammed releasing stimuli.

Flagellation The practice of submitting to whipping for sexual or penitential purposes.

Flagellomania Sexual excitement aroused by whipping.

Flagellum A whiplike part or process of some cells, especially of certain bacteria, protozoans, and so on, that is an organ of locomotion or produces a current in the surrounding fluid.

Flashbulb memory Important events that are fixed mainly in visual and auditory memory.

Flooding A form of behaviour therapy based on concepts derived from classical conditioning in which the patient exposes herself to whatever she is afraid of, thus extinguishing her fear.

Flow chart 1. In computer science, a diagram that shows the step-by-step operation of a computer program.

2. In human cognition, similar diagrams that show the hypothesized flow of information as it is thought to be processed by the human mind.

Fluid and crystallized intelligence In psychometrics, fluid and crystallized intelligence (abbreviated gf and gc respectively) are factors of intelligence test scores originally described by Raymond Cattell. Crystallized intelligence is usually described as being dependent on learning, while fluid intelligence is independent of past experience.

Fluid intelligence One of the two most important general factors influencing human performance, which relates to a person's capacity to solve new problems and to learn.

Flynn effect Phenomenon showing that people's performance on IQ tests has improved over time in industrialized countries.

Folie a deux The occurrence of psychosis in two persons who are closely associated to each other (i.e., husband and wife).

Folkway Any way of thinking, feeling, behaving, and so on common to members of the same social group.

Foot-in-the-door technique A technique of persuasion, initially used by door-to-door salespeople, in which one first obtains a small concession that then makes it easier to persuade the target to make a subsequent, larger concession.

Force Necessary to the preservation of a society because people have innate aggressive, incestuous, and cannibalistic tendencies and won't work unless coerced. Freud regarded the conflict between person and culture to be, at bottom, unresolvable; at best, society's dictates and codes become internalized into a **superego** that acts as an internal agent of society.

Forced compliance effect An individual forced to act or speak publicly in a manner contrary to his own beliefs may change his own views in the direction of the public action. But this will happen only if his reward for the false public pronouncement is relatively small. If the reward is large, there is no dissonance and hence no attitude change.

Forced fantasy An emotional fantasy that is deliberately promoted by the therapist (analyst). This procedure has been criticized for its ability to create false memories.

Forebrain The biggest and most complex part of the brain, which includes structures such as the thalamus, the hypothalamus, the limbic system, and the cerebrum.

Foreclosure An individual's premature establishment of an identity based on parental choices rather than their own.

Forensic Pertaining to the courts.

Forensic anthropology The forensic anthropologist examines the victims bones to determine a number of key facts. Information such as gender, age, looks, previous trauma, and disease can all be found. The forensic anthropologist is often of key relevance to the identification of remains. They use a number of means, including molecular DNA analysis.

Forensic artist The forensic artist provides an elaborate sketch of the offender. This process is undertaken via the information from an eyewitness. Many investigators now use computer programs to develop offender renditions.

Forensic chemistry The forensic chemist studies the molecular aspects of the crime scene. They can match fibres, paint, and dyes to particular objects. They will identify relevant chemicals and particles.

Forensic dentistry These experts serve a identification function. Via the analysis of a corpses teeth and previous dental records they can make a positive identification. They also will analyze bite patterns so that they can identify who was eating a particular meal or even who bit somebody.

Forensic entomology The entomologist studies insects. A number of pieces of key information can be discovered from this analysis. The rate of body decay via insects can directly correlate to time of death. The presence of certain insects can identify previous dump sites, etc.

Forensic geology The forensic geologist can determine where a person or object has been by analyzing soil samples. Soil can be found on a pair of shoes, tire treads, or a body. They can be matched up with common soil types to predict quite specifically where the object/person has been.

Forensic linguistics The forensic linguist analyses either the spoken or written word. They can identify whether a message was presented by the same individual, what the individuals underlying intent is, the individuals educational and cultural background, as well as the presence of pathology.

Forensic odontology see 'forensic dentistry'.

Forensic pathology The forensic pathologist analyzes the remains of a body. They attempt to determine the cause and time of death via autopsy.

Forensic photography The crime scene photographer attempts to record every component of the crime scene via photograph. They depict the scene from multiple angles, using multi-functional cameras, and through the consistent evaluation of size and distance.

Forensic psychologist A Ph.D. psychologist whom specializes in the realm of forensics. The forensic psychologist may have either a Ph.D. in forensic psychology, clinical psychology with a forensic focus, or experimental psychology with a forensic focus.

Forensic psychology The discipline of psychology that deals with the legal system, including the front end operations (police work, investigation), legal proceedings (expert witness, competency & psychological assessment, jury selection), and institutionalization (confinement, treatment, parole hearings). Forensic psychologists may also be called on to: evaluate new laws and programs, assist in the assessment and hiring of new police officers, etc.

Forensic sculptors Like the forensic artist, the sculptor attempts to create an image of either an offender or a victim. The sculptor's goal is to create a three dimensional version of the image.

Forensic serology The serologist studies blood and other bodily fluids for identification purposes. The serologist is often involved in DNA fingerprinting (the identification of an individual based on body cells).

Forgetting curve A curve showing the inverse relationship between memory and the retention interval.

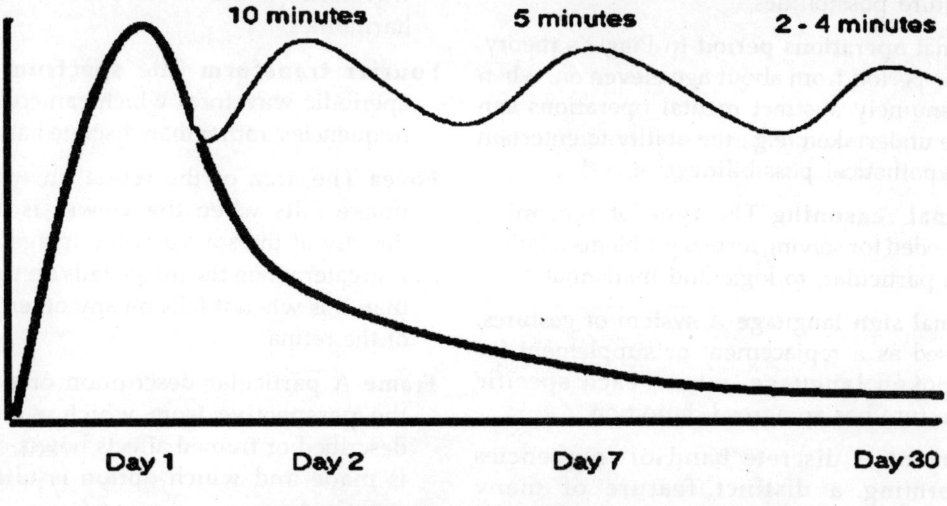

Fig. Forgetting curve

Form constancy The tendency to perceive objects as having a constant form, even when they are rotated or their distance from the observer changes.

Form stream The neural pathway in the visual cortex thought to be concerned primarily with the analysis of the stimulus's spatial structure.

Formal assessment The systematic procedures and measurement instruments used by trained professionals to assess an individual's functioning, aptitudes, abilities, or mental states.

Formal concept A category of objects or situations defined by listing their common essential characteristics, as dictionary definitions do.

Formal operational stage Pavlov's fourth and final stage of cognitive development where thinking becomes more abstract.

Formal operational thinking A term defined by jean piaget to describe the stage, arrived at by around age 11 or 12, when children are able to first think in abstract, logical and systematic ways.

Formal operational thought Deals abstractly with hypothetical situations and reason.

Formal operations According to Piaget, individuals enter this stage in adolescence as they gain the ability to classify and compare objects and ideas, systematically seek solutions to problems, and consider future possibilities.

Formal operations period In Piaget's theory, the period from about age eleven on, when genuinely abstract mental operations can be undertaken (e.g., the ability to entertain hypothetical possibilities).

Formal reasoning The type of reasoning needed for solving formal problems relating, in particular, to logic and mathematics.

Formal sign language A system of gestures, used as a replacement or supplement for spoken language, where each specific gesture has an agreed definition.

Formant A discrete band of frequencies forming a distinct feature of many (especially periodic) speech sounds, for example vowel sounds.

Formant transition The changes over time of the frequencies of the formants in some speech sounds, for example, voiced stops (/d/, /g/, /b/).

Formative evaluation Tests or assessments administered during units of instruction that measure progress and guide the content and pace of lessons.

Formative quiz Evaluation designed to determine whether additional instruction is needed.

Formats A type of familiar play or care routine within which carers constructively support a child's language development.

Forward pairing A classical conditioning procedure in which the conditioned stimulus (CS) precedes the unconditioned stimulus (US). This contrasts with simultaneous pairing, in which CS and US are presented simulataneously, and backward pairing, in which CS follows US.

Foundational theories Frameworks for initial understanding formulated by children to explain their experiences of the world.

Fourier series The spectrum of a periodic waveform, which is composed of discrete harmonics.

Fourier transform The spectrum of an aperiodic waveform which can contain all frequencies, rather than discrete harmonics.

Fovea The area of the retina on which an image falls when the viewer is looking directly at the source of the image. Acuity is greater when the image falls on the fovea than it is when it falls on any other portion of the retina.

Frame A particular description of a choice; the perspective from which a choice is described or framed affects how a decision is made and which option is ultimately exercised.

Frames of reference The developing child's expand-ing range of geographical experience, labelled as edocentric (relating to body-part coordination), allocentric (relating to the immediate world) and geocentric (relating to the spatial relationships between places).

Framing A heuristic that affects the subjective desirability of an event by changing the standard of reference for judging the desirability of that event.

Francophobia Fear of France or French culture. (Gallophobia, Galiophobia).

Fraternal twins Twins that arise from two different eggs that are (simultaneously) fertilized by different sperm cells. Their genetic similarity is no greater than that between ordinary siblings.

Fraudulent witness While not having any first hand knowledge of the crime, the fraudulent witness is an active attention seeker whom comes forward to offer fraudulent evidence.

Free association A psychoanalytic procedure in which the client is encouraged to speak freely, without censoring possibly embarrassing or socially unacceptable thoughts or ideas.

Free nerve ending An unencapsulated (naked) dendrite of somatosensory neurons.

Free recall A test of memory that asks for as many items in a list as a research participant can recall regardless of order.

Free recall learning A task requiring recall of a list of items in any order.

Free riders People who benefit from the group but give little in return.

Frequency 1. In sound waves or light waves, the number of wave peaks per second. In sound, frequency governs the perceived pitch of the sound; in light, frequency governs the perceived hue of the light.

2. In statistical analysis, the number of occurrences of a particular observation.

Frequency distribution A summary of how frequently each score appears in a set of observations.

Frequency effect The phenomenon in memory which states that we tend to remember information better if it is repeated.

Frequency polygon The resulting shape from finding the midpoints in a frequency distribution (histogram) and then connecting the points with a line.

Frequency theory A proposal for how auditory frequency is encoded by the nervous system. According to the theory, firing rate in the auditory nerve corresponds to the frequency of the incoming sound wave; the nervous system then interprets different firing rates as different pitches. This theory is probably correct for low frequencies, but not for higher ones.

Freud, Anna (1895 - 1982) It seems that every time Freud felt he had his successor picked out, the nominee would abandon him. At least, that's what happened with Jung and Adler. In the meantime, though, his daughter Anna was attending lectures, going through analysis with her father, and generally moving towards a career as a lay psychoanalyst. She also became his caretaker after he developed cancer in 1923. She became at very least her father's symbolic successor.

Unlike Jung and Adler, she remained faithful to the basic ideas her father developed. However, she was more interested in the dynamics of the psyche than in its structure, and was particularly fascinated by the place of the ego in all this. Freud had, after all, spent most of his efforts on the id and the unconscious side of psychic life. As she rightly pointed out, the ego is the "seat of observation" from which we observe the work of the id and the superego and the unconscious generally, and deserves study in its own right.

She is probably best known for her book *The Ego and the Mechanisms of Defense,* in which she gives a particularly clear description of how the defenses work, including some special attention to adolescents' use of defenses. The defenses section of the chapter on Freud in this text is based as much on Anna's work as on Sigmund's.

This focus on the ego began a movement in psychoanalytic circles called ego psychology that today represents, arguably, the majority of Freudians. It takes Freud's earlier work as a crucial foundation, but extends it into the more ordinary, practical, day-to-day world of the ego. In this way, Freudian theory can be applied, not only to psychopathology, but to social and developmental issues as well. Erik Erikson is the best-known example of an ego psychologist.

But Anna Freud was not primarily a theoretician. Her interests were more practical, and most of her energies were devoted to the analysis of children and adolescents, and to improving that analysis. Her father, after all, had focused entirely on adult patients. Although he wrote a great deal about development, it was from the perspectives of these adults. What do you do with the child, for whom family crises and traumas and fixations are present events, not dim recollections?.

First, the relationship of the child to the therapist is different. The child's parents are still very much a part of his or her life, a part the therapist cannot and should not try to usurp. But neither can the therapist pretend to be just another child rather than an authority figure. Anna Freud found that the best way to deal with this "transference problem" was the way that came most naturally: be a caring adult, not a new playmate, not a substitute parent. Her approach seems authoritarian by the standards of many modern child therapies, but it might make more sense.

Another problem with analyzing children is that their symbolic abilities are not as advanced as those of adults. The younger ones, certainly, may have trouble relating their emotional difficulties verbally. Even older children are less likely than adults to bury their problems under complex symbols. After all, the child's problems are here-and-now; there hasn't been much time to build up defenses. So the problems are close to the surface and tend to be expressed in more direct, less symbolic, behavioral and emotional terms.

Most of her contributions to the study of personality come out of her work at the Hamstead Child Therapy Clinic in London, which she helped to set up. Here, she found that one of the biggest problems was communications among therapists: Whereas adult problems were commu-nicated by means of traditional labels, children's problems could not be.

Because children's problems are more immediate, she reconceptualized them in terms of the child's movement along a developmental time-line. A child keeping pace with most of his or her peers in terms of eating behaviors, personal hygiene, play styles, relationships with other children, and so on, could be considered healthy. When one aspect or another of a child's development seriously lagged behind the rest, the clinician could assume that there was a problem, and could communicate the problem by describing the particular lag.

Freud, Sigmund (1856-1939) Freud was born in Frieberg, Moravia in 1856, but when he was four years old his family moved to Vienna, where Freud was to live and work until the last year of his life. In 1937 the Nazis annexed Austria, and Freud, who was Jewish, was allowed to leave for England. For these reasons, it was above all with the city of Vienna that Freud's name was destined to be deeply associated for posterity, founding as he did what was to become known as the 'first Viennese school' of psychoanalysis, from which, it is fair to say, psychoanalysis as a movement and all subsequent developments in this field flowed. The scope of Freud's interests, and of his professional training, was very broad - he always considered himself first and foremost a scientist, endeavouring to extend the compass of human knowledge, and to this end (rather than to the practice of medicine) he enrolled at the medical school at the University of Vienna in 1873. He concentrated initially on biology, doing research in physiology for six years under the great German scientist Ernst Brücke, who was director of the Physiology Laboratory at the University, thereafter specialising in neurology. He received his medical degree in 1881, and having become engaged to be married in 1882, he rather reluctantly took up more secure and financially rewarding work as a doctor at Vienna General Hospital. Shortly after his marriage in 1886 - which was extremely happy, and gave Freud six children, the youngest of whom, Anna, was herself to become a distinguished psychoanalyst - Freud set up a private practice in the treatment of psychological disorders, which gave him much of the clinical material on which he based his theories and his pioneering techniques. In 1885-86 Freud spent the greater part of a year in Paris, where he was deeply impressed by the work of the French neurologist Jean Charcot, who was at that time using hypnotism to treat hysteria and other abnormal mental conditions. When he returned to Vienna, Freud experimented with hypnosis, but found that its beneficial effects did not last. At this point he decided to adopt instead a method suggested by the work of an older Viennese colleague and friend, Josef Breuer, who had discovered that when he encouraged a hysterical patient to talk uninhibitedly about the earliest occurrences of the symptoms, the latter sometimes gradually abated. Working with Breuer, Freud formulated and developed the idea that many neuroses (phobias, hysterical paralyses and pains, some forms of paranoia, etc.) had their origins in deeply traumatic experiences which had occurred in the past life of the patient but which were now forgotten, hidden from consciousness; the treatment was to enable the patient to recall the experience to consciousness, to confront it in a deep way both intellectually and emotionally, and in thus discharging it, to remove the underlying psychological causes of the neurotic symptoms. This technique, and the theory from which it is derived, was given its classical expression in *Studies in Hysteria*, jointly published by Freud and Breuer in 1895. Shortly thereafter, however, Breuer, found that he could not agree with what he regarded as the excessive emphasis which Freud placed upon the sexual origins and content of neuroses, and the two parted company, with Freud continuing to work alone to develop and refine the theory and practice of psychoanalysis. In 1900, after a protracted period of self-analysis, he published *The Interpretation of Dreams*, which is generally regarded as his greatest work, and this was followed in 1901 by *The Psychopathology of Everyday Life*, and in 1905 by *Three Essays on the Theory of Sexuality*. Freud's psychoanalytic theory was initially not well received - when its existence was acknowledged at all it was usually by people who were, as Breuer had foreseen, scandalised by the emphasis placed on sexuality by Freud - and it was not until 1908, when the first International

Psychoanalytical Congress was held at Salzburg, that Freud's importance began to be generally recognised. This was greatly facilitated in 1909, when he was invited to give a course of lectures in the United States, which were to form the basis of his 1916 book *Five Lectures on Psycho-Analysis.* From this point on Freud's reputation and fame grew enormously, and he continued to write prolifically until his death, producing in all more than twenty volumes of theoretical works and clinical studies. He was also not adverse to critically revising his views, or to making fundamental alterations to his most basic principles when he considered that the scientific evidence demanded it - this was most clearly evidenced by his advancement of a completely new tripartite (*id, ego,* and *super-ego*) model of the mind in his 1923 work *The Ego and the Id.* He was initially greatly heartened by attracting followers of the intellectual calibre of Adler and Jung, and was correspondingly disappointed personally when they both went on to found rival schools of psychoanalysis - thus giving rise to the first two of many schisms in the movement - but he knew that such disagreement over basic principles had been part of the early development of every new science. After a life of remarkable vigour and creative productivity, he died of cancer while exiled in England in 1939.

Although a highly original thinker, Freud was also deeply influenced by a number diverse factors which overlapped and interconnected with each other to shape the development of his thought. As indicated above, both Charcot and Breuer had a direct and immediate impact upon him, but some of the other factors, though no less important than these, were of a rather different nature. First of all, Freud himself was very much a Freudian - his father had two sons by a previous marriage, Emmanuel and Philip, and the young Freud often played with Philip's son John, who was his own age. Freud's own self-analysis - which forms the core of his masterpiece *The Interpretation of Dreams* - originated in the emotional crisis which he suffered on the death of his father, and the series of dreams to which this gave rise. This analysis revealed to him that the love and admiration which he had felt for his father were mixed with very contrasting feelings of shame and hate (such a mixed attitude he termed 'ambivalence'). Particularly revealing was his discovery that he had often fantasised as a youth that his half-brother Philip (who was of an age with his mother) was really his father, and certain other signs convinced him of the deep underlying meaning of this fantasy - that he had wished his real father dead, because he was his rival for his mother's affections. This was to become the personal (though by no means exclusive) basis for his theory of the Oedipus complex.

Secondly, and at a more general level, account must be taken of the contemporary scientific climate in which Freud lived and worked. In most respects, the towering scientific figure of nineteenth century science was Charles Darwin, who had published his revolutionary *Origin of Species* when Freud was four years old. The evolutionary doctrine radically altered the prevailing conception of man - whereas before man had been seen as a being different in nature to the members of the animal kingdom by virtue of his possession of an immortal soul, he was now seen as being part of the natural order, different from non-human animals only in degree of structural complexity. This made it possible and plausible, for the first time, to treat man as an object of scientific investigation, and to conceive of the vast and varied range of human behaviour, and the motivational causes from which it springs, as being amenable in principle to scientific explanation. Much of the creative work done in a whole variety of diverse scientific fields over the next century was to be inspired by, and derive sustenance from, this new world-view, which Freud, with his enormous esteem for science, accepted implicitly.

An even more important influence on Freud, however, came from the field of physics. The second 50 years of the nineteenth century saw monumental advances in contemporary physics, which were largely initiated by the formulation of the principle of the conservation of energy by Helmholz. This principle states, in effect, that the total amount of energy in any given physical system is always constant, that energy quanta can be changed but not annihilated, and consequently that when energy is moved from one part of the system it must reappear in another part. The progressive application of this principle led to the monumental discoveries in the fields of thermodynamics, electromagne-ticism, and nuclear physics which, with their associated technologies, have so comprehen-sively transformed the contemporary world. As we have seen, when he first came to the University of Vienna Freud worked under the direction of Ernst Brücke, who in 1874 published a book setting out the view that all living organisms, including the human one, are essentially energy-systems to which, no less than to inanimate objects, the principle of the conservation of energy applies. Freud, who had great admiration and respect for Brücke, quickly adopted this new 'dynamic physiology' with enthusiasm. From there it was but a short conceptual step - but one which Freud was the first to take, and on which his claim to fame is largely grounded - to the view that there is such a thing as 'psychic energy', that the *human personality* is also an energy-system, and that it is the function of psychology to investigate the modifications, transmissions, and conversions of 'psychic energy' within the personality which shape and determine it. This latter conception is the very cornerstone of Freud's psycho-analytic theory.

Freud's theory of the unconscious, then, is highly deterministic, a fact which, given the nature of nineteenth century science, should not be surprising. Freud was arguably the first thinker to apply deterministic principles systematically to the sphere of the mental, and to hold that the broad spectrum of human behaviour is explicable only in terms of the (usually hidden) mental processes or states which determine it. Thus, instead of treating the behaviour of the neurotic as being causally inexplicable - which had been the prevailing approach for centuries - Freud insisted, on the contrary, on treating it as behaviour for which is meaningful to seek *an explanation* by searching for *causes* in terms of the mental states of the individual concerned. Hence the significance which he attributed to slips of the tongue or pen, obsessive behaviour, and dreams - all, he held, are determined by hidden causes in the person's mind, and so they reveal in covert form what would otherwise not be known at all. This suggests the view that freedom of the will is, if not completely an illusion, certainly more tightly circu-mscribed than is commonly believed, for it follows from this that whenever we make a choice we are governed by hidden mental processes of which we are unaware and over which we have no control. The postulate that there *are* such things as unconscious mental states at all is a direct function of Freud's determinism, his reasoning here being simply that the principle of causality *requires* that such mental states should exist, for it is evident that there is frequently nothing in the conscious mind which can be said to cause neurotic or other behaviour. An 'unconscious' mental process or event, for Freud, is not one which merely happens to be out of consciousness at a given time, but is rather one which *cannot*, except through protracted psychoanalysis, be brought to the forefront of consciousness. The postulation of such unconscious mental states entails, of course, that the mind is not, and cannot be, identified with consciousness or that which can be an object of consciousness - to employ a much-used analogy, it is rather structurally akin to an iceberg, the bulk of it lying below the surface, exerting a dynamic and determining influence upon the part which is amenable to direct inspection, the conscious mind.

Deeply associated with this view of the mind is Freud's account of the instincts or drives. The instincts, for Freud, are the principal motivating forces in the mental realm, and as such they 'energise' the mind in all of its functions. There are, he held, an indefinitely large number of such instincts, but these can be reduced to a small number of basic ones, which he grouped into two broad generic categories, *Eros* (the life instinct), which covers all the self-preserving and erotic instincts, and *Thanatos* (the death instinct), which covers all the instincts towards aggression, self-destruction, and cruelty. Thus it is a mistake to interpret Freud as asserting that *all* human actions spring from motivations which are sexual in their origin, since those which derive from Thanatos are not sexually motivated - indeed, Thanatos is the irrational urge to destroy the source of all sexual energy in the annihilation of the self. Having said that, it is undeniably true that Freud gave sexual drives an importance and centrality in human life, human actions, and human behaviour which was new (and to many, shocking), arguing as he does both that the sexual drives exist and can be discerned in children from birth (the theory of infantile sexuality), and that sexual energy (*libido*) is the single most important motivating force in adult life. However, even here a crucial qualification has to be added - Freud effectively redefined the term 'sexuality' here to make it cover *any form of pleasure* which is or can be derived from the body. Thus his theory of the instincts or drives is essentially that the human being is energised or driven from birth by the desire to acquire and enhance bodily pleasure.

Freud's theory of infantile sexuality must be seen as an integral part of a broader developmental theory of human personality. This had its origins in, and was a generalisation of, Breuer's earlier discovery that traumatic childhood events could have devastating negative effects upon the adult individual, and took the form of the general thesis that early childhood sexual experiences were the crucial factors in the determination of the adult personality. From his account of the instincts or drives it followed that from the moment of birth the infant is driven in his actions by the desire for bodily/sexual pleasure, where this is seen by Freud in almost mechanical terms as the desire to release mental energy. Initially, infants gain such release, and derive such pleasure, through the act of sucking, and Freud accordingly terms this the 'oral' stage of development. This is followed by a stage in which the locus of pleasure or energy release is the anus, particularly in the act of defecation, and this is accordingly termed the 'anal' stage. Then the young child develops an interest in its sexual organs as a site of pleasure (the 'phallic' stage), and develops a deep sexual attraction for the parent of the opposite sex, and a hatred of the parent of the same sex (the 'Oedipus complex'). This, however, gives rise to (socially derived) feelings of guilt in the child, who recognises that it can never supplant the stronger parent. In the case of a male, it also puts the child at risk, which he perceives - if he persists in pursuing the sexual attraction for his mother, he may be harmed by the father; specifically, he comes to fear that he may be castrated. This is termed '*castration anxiety*'. Both the attraction for the mother and the hatred are usually repressed, and the child usually resolves the conflict of the Oedipus complex by coming to identify with the parent of the same sex. This happens at the age of five, whereupon the child enters a 'latency' period, in which sexual motivations become much less pronounced. This lasts until puberty, when mature genital development begins, and the pleasure drive refocuses around the genital area.

This, Freud believed, is the sequence or progression implicit in *normal* human development, and it is to be observed that at the infant level the instinctual attempts to satisfy the pleasure drive are frequently checked by parental control and social coercion. The developmental process, then, is for the child essentially a movement through a series of *conflicts*, the successful resolution of which is crucial to adult mental health. Many mental illnesses, particularly hysteria, Freud held, can be traced back to unresolved conflicts experienced at this stage, or to events which otherwise disrupt the normal pattern of infantile development. For example, homosexuality is seen by some Freudians as resulting from a failure to resolve the conflicts of the Oedipus complex, particularly a failure to identify with the parent of the same sex; the obsessive concern with washing and personal hygiene which characterises the behaviour of some neurotics is seen as resulting from unresolved conflicts/repressions occurring at the anal stage.Freud's account of the unconscious, and the psychoanalytic therapy associated with it, is best illustrated by his famous tripartite model of the structure of the mind or personality (although, as we have seen, he did not formulate this until 1923), which has many points of similarity with the account of the mind offered by Plato over 2,000 years earlier. The theory is termed 'tripartite' simply because, again like Plato, Freud distinguished three structural elements within the mind, which he called *id*, *ego*, and *super-ego*. The *id* is that part of the mind in which are situated the instinctual sexual drives which require satisfaction; the *super-ego* is that part which contains the 'conscience', viz. socially-acquired control mechanisms (usually imparted in the first instance by the parents) which have been internalised; while the *ego* is the conscious self created by the dynamic tensions and interactions between the *id* and the *super-ego*, which has the task of reconciling their conflicting demands with the requirements of external reality. It is in this sense that the mind is to be understood as a dynamic energy-system. All objects of consciousness reside in the *ego*, the contents of the *id* belong permanently to the unconscious mind, while the *super-ego* is an unconscious screening-mechanism which seeks to limit the blind pleasure-seeking drives of the *id* by the imposition of restrictive rules. There is some debate as to how literally Freud intended this model to be taken (he appears to have taken it extremely literally himself), but it is important to note that what is being offered here is indeed *a theoretical model*, rather than a description of an observable object, which functions as a frame of reference to explain the link between early childhood experience and the mature adult (normal or dysfunctional) personality. Freud also followed Plato in his account of the nature of mental health or psychological well-being, which he saw as the establishment of a harmonious relationship between the three elements which constitute the mind. If the external world offers no scope for the satisfaction of the *id's* pleasure drives, or, more commonly, if the satisfaction of some or all of these drives would indeed transgress the moral sanctions laid down by the *super-ego*, then an inner conflict occurs in the mind between its constituent parts or elements - failure to resolve this can lead to later neurosis. A key concept introduced here by Freud is that the mind possesses a number of 'defence mechanisms' to attempt to prevent conflicts from becoming too acute, such as *repression* (pushing conflicts back into the unconscious), *sublimation* (channelling the sexual drives into the achievement socially acceptable goals, in art, science, poetry, etc.), *fixation* (the failure to progress beyond one of the developmental stages), and *regression* (a return to the behaviour characteristic of one of the stages). Of these, repression is the most important, and Freud's account of this is as follows: when a person experiences an instinctual impulse to behave in a manner which the *super-ego* deems to be reprehensible (e.g. a strong erotic impulse on the part of the child towards the parent of the opposite sex), then it is possible for the mind push it away, to *repress* it into the unconscious. Repression is thus one of the central defence mechanisms by which the ego seeks to avoid internal conflict and pain, and to reconcile reality with the demands of both *id* and *super-ego*. As such it is completely normal and an integral part of the developmental process through which every child must pass on the way to adulthood. However, the repressed instinctual drive, as an energy-form, is not and cannot be *destroyed* when it is repressed - it continues to exist intact in the unconscious, from where it exerts a determining force upon the conscious mind, and can give rise to the dysfunctional behaviour characteristic of neuroses. This is one reason why dreams and slips of the tongue possess such a strong symbolic significance for Freud, and why their analysis became such a key part of his treatment - they represent instances in which the vigilance of the *super-ego* is relaxed, and when the repressed drives are accordingly able to present themselves to the conscious mind in a transmuted form. The difference between 'normal' repression and the kind of repression which results in neurotic illness is one of *degree*, not of kind - the compulsive behaviour of the neurotic is itself a behavioural manifestation of an instinctual drive repressed in childhood. Such behavioural symptoms are highly irrational (and may even be perceived as such by the neurotic), but are completely beyond the control of the subject, because they are driven by the now unconscious repressed impulse. Freud positioned the key repressions, for both the normal individual and the neurotic, in the first five years of childhood, and, of course, held them to be essentially sexual in nature - as we have seen, repressions which disrupt the process of infantile sexual development in particular, he held, lead to a strong tendency to later neurosis in adult life. The task of psychoanalysis as a therapy is to find the repressions which are causing the neurotic symptoms by delving into the unconscious mind of the subject, and by bringing them to the forefront of consciousness, to allow the ego to confront them directly and thus to discharge them.

Freud's account of the sexual genesis and nature of neuroses led him naturally to develop a clinical treatment for treating such disorders. This has become so influential today that when people speak of 'psychoanalysis' they frequently refer exclusively to the clinical treatment; however, the term properly designates both the clinical treatment and the theory which underlies it. The aim of the method may be stated simply in general terms - to re-establish a harmonious relationship between the three elements which constitute the mind by excavating and resolving unconscious repressed conflicts. The actual method of treatment pioneered by Freud grew out of Breuer's earlier discovery, mentioned above, that when a hysterical patient was encouraged to talk freely about the earliest occurrences of her symptoms and fantasies, the symptoms began to abate, and were eliminated entirely she was induced to remember the initial trauma which occasioned them. Turning away from his early attempts to explore the unconscious through hypnosis, Freud further developed this 'talking cure', acting on the assumption that the repressed conflicts were buried in the deepest recesses of the unconscious mind. Accordingly, he got his patients to relax in a position in which they were deprived of strong sensory stimulation, even of keen awareness of the presence of the analyst (hence the famous use of the couch, with the analyst virtually silent and out of sight), and then encouraged them to speak freely and uninhibitedly, preferably without forethought, in the belief that he could thereby discern the unconscious forces lying behind what was said. This is the method of 'free-association', the rationale for which is similar to that involved in the analysis of dreams - in both cases the *super-ego* is to some degree disarmed, its efficiency as a screening mechanism is moderated, and material is allowed to filter through to the conscious *ego* which would otherwise be completely repressed. The process is necessarily a difficult and protracted one, and it is therefore one of the primary tasks of the analyst to help the patient to recognise, and to overcome, his own natural resistances, which may exhibit themselves as hostility towards the analyst. However, Freud always took the occurrence of resistance as a sign that he was on the right track in his assessment of the underlying unconscious causes of the patient's condition. The patient's dreams are of particular interest, for reasons which we have already partly seen. Taking it that the *super-ego* functioned less effectively in sleep, as in free association, Freud made a distinction between the *manifest* content of a dream (what the dream appeared to be about on the surface) and its *latent* content (the unconscious, repressed desires or wishes which are its real object). The correct interpretation of the patient's dreams, slips of tongue, free-associations, and responses to carefully selected questions leads the analyst to a point where he can locate the unconscious repressions producing the neurotic symptoms, invariably in terms of the patient's passage through the sexual developmental process, the manner in which the conflicts implicit in this process were handled, and the libidinal content of his family relationships. To effect a cure, he must facilitate the patient himself to become conscious of unresolved conflicts buried in the deep recesses of the unconscious mind, and to confront and engage with them directly.

In this sense, then, the object of psychoanalytic treatment may be said to be a form of self-understanding - once this is acquired, it is largely up to the patient, in consultation with the analyst, to determine how he shall handle this newly-acquired understanding of the unconscious forces which motivate him. One possibility, mentioned above, is the channelling of the sexual energy into the achievement of social, artistic or scientific goals - this is sublimation, which Freud saw as the motivating force behind most great cultural achievements. Another would be the conscious, rational control of the formerly repressed drives - this is suppression. Yet another would be the decision that it is the *super-ego*, and the social constraints which inform it, which are at fault, in which case the patient may decide in the end to satisfy the instinctual drives. But in all cases the cure is effected essentially by a kind of catharsis or purgation - a *release* of the pent-up psychic energy, the constriction of which was the basic cause of the neurotic illness.

This is a crucially important issue, since Freud not alone saw himself first and foremost as a pioneering scientist, but repeatedly asserted that the significance of psychoanalysis is that it is a *new science*, incorporating a new scientific method of dealing with the mind and with mental illness. And there can be no doubt but that this has been the chief attraction of the theory for most of its advocates since then - on the face of it, it has the appearance of being, not just a scientific theory, but an enormously *strong* scientific theory, with the capacity to accommodate, and explain, every possible form of human behaviour. However, it is precisely this latter which, for many commentators, undermines its claim to scientific status. On the question of what makes a theory a genuinely scientific one, Karl Popper's criterion of demarcation, as it is called, has now gained very general acceptance: viz. that every genuine scientific theory must be testable, and therefore *falsifiable*, at least in principle - in other words, if a theory is incompatible with *possible* observations it is scientific; conversely, a theory which is compatible with *all* possible observations is unscientific. Thus the principle of the conservation of energy, which influenced Freud so greatly, is a scientific one, because it is falsifiable - the discovery of a physical system in which the total amount of energy was not constant would conclusively show it to be false. And it is argued that nothing of the kind is possible with respect to Freud's theory - if, in relation to it, the question is asked: 'What does this theory imply which, if false, would show the whole theory to be false?', the answer is 'nothing', the theory is compatible with *every possible* state of affairs - it cannot be falsified by anything, since it purports to explain everything. Hence it is concluded that the theory is not scientific, and while this does not, as some critics claim, rob it of all value, it certainly diminishes its intellectual status, as that was and is projected by its strongest advocates, including Freud himself.

A related (but perhaps more serious) point is that the coherence of the theory is, at the very least, questionable. What is attractive about the theory, even to the layman, is that it seems to offer us long sought-after, and much needed, *causal explanations* for conditions which have been a source of a great deal of human misery. The thesis that neuroses are caused by unconscious conflicts buried deep in the unconscious mind in the form of repressed libidinal energy would appear to offer us, at last, an insight in the causal mechanism underlying these abnormal psychological conditions as they are expressed in human behaviour, and further show us how they are related to the psychology of the 'normal' person. However, even this is questionable, and is a matter of much dispute. In general, when it is said that an event X causes another event Y to happen, both X and Y are, and must be, *independently identifiable*. It is true that this is not always a simple process, as in science causes are sometimes unobservable (sub-atomic particles, radio and electromagnetic waves, molecular structures, etc.), but in these latter cases there are clear 'correspondence rules' connecting the unobservable causes with observable phenomena. The difficulty with Freud's theory is that it offers us entities (repressed unconscious conflicts, for example) which are said to be the unobservable causes of certain forms of behaviour, but there are *no* correspondence rules for these alleged causes - they cannot be identified *except* by reference to the behaviour which they are said to cause (i.e. the analyst does not demonstratively assert: '*This* is the unconscious cause, and *that* is its behavioural effect'; he asserts: '*This* is the behaviour, therefore its unconscious cause *must* exist'). And this does raise serious doubts as to whether Freud's theory offers us genuine causal explanations at all.

Freud, some theoretical differences 1. Subjective finalism vs reductive materialism. For Freud, religion = father complex.

2. Libido as neutral psychic energy. Less emphasis on sexuality.

3. Incest fantasies seen as gropings for unity and manifestations of kinship libido. Union of opposites. Urge to return to childhood. Only adults are capable of incest because of their formed sexuality, which regresses.

4. Sublimation discredited because religious and cultural activities are authentic parts of psychological existence and manifestations of an autonomous instinct to create, not derivatives of sexual impulses. Jung posited instead a transformation of psychic energy. Jung compared "sublimation" to the alchemical trick of turning lead into gold—"Unfortunately, the secret of converting energy without the consumption of a still greater quantity of energy has never yet been discovered by the physicists" ("Sigmund Freud In His Historical Setting").

5. Reductive therapy is useful mostly for people in the first half of life.

6. Id: the "red" or lower end of the psychic spectrum rather than the source of all psychic activity.

7. Superego: smuggles in the image of Jehovah in the dress of a theory. Similar to Jung's "collective consciousness." Sum of all the collective beliefs, ideas, and values consciously handed down by tradition and undergirded by the primordial images. Also, the Self as long as it's unconscious and therefore projected onto public opinion.

8. Drive theory: too soon to say what's biological and what's psychological.

9. The unconscious isn't a thing, it's a process.

10. We cure in spite of transference, not because of it.

11. Nonpersonal forces aren't just "archaic vestiges.".

12. Death drive is just a play of opposites. Libido has two instincts or urges: to live and to die. The death part = incestuous libido harking back to unconsciousness, womb, instinct.

13. The unconscious isn't just an appendix of consciousness. It also contains future developments and the archetypes.

14. Dreams don't disguise (because nature doesn't); they are factual pictures that speak in a symbolic language our rationalistic minds have trouble with. The manifest disguise is just our lack of understanding of symbolic language. A dream can't produce a clear thought—that's the ego's job. Nor do they usually exhibit fixed symbols. Also, only some dreams are wish-fulfillments.

15. The censor: for Jung, the exclusive directedness of conscious contents that excludes an unconscious content.

16. Regression of libido isn't just incest, infantile sexuality, fixation, etc., but a vital process.

17. Children aren't polymorphous-perverse, but polyvalent.

18. The Oedipus complex isn't a cause of illness, it's a symptom. Wanting mom = regressive longing for her.

19. Latency is actually the true beginnings of sexuality.

20. It's not just a matter of getting in touch with the instincts, but also with the archetypes that give them form, meaning, and bounds.

Freud's theory of dreams A theory that holds that at bottom all dreams are attempts to fulfill a wish. The wish fulfillment is in the latent dream, which represents the sleeper's hidden desires. This latent dream is censored and reinterpreted to avoid anxiety. It reemerges in more acceptable form as the manifest dream, the dream the sleeper remembers upon awakening.

Freudian slip A mistake, usually of speech, made due to the collision of conscious and unconscious conflicts. An example made famous by Freud: the reluctant official who opens a meeting by accidentally declaring it closed, thereby revealing his true wish.

Friendship A loving relationship characterized by intimacy, but not by passion or commitment.

Frigidity Female impotence, lack of sexual response in females.

Frigophobia Fear of cold or cold things. (Cheimaphobia, Cheimatophobia, Psychrophobia).

Frontal lobe The front portion of the cerebral cortex, including Broca's speech area and the motor cortex; damage impairs movement, planning, and flexibility in behavioural strategies.

Frottage Sexual gratification achieved by rubbing against the clothing of a member of the opposite sex in a crowd.

Frustration The feelings, thoughts, and behaviours associated with not achieving a particular goal or the belief that a goal has been prematurely interrupted.

Frustration agression hypothesis The drive theory of aggression that suggests frustration builds and creates aggression (ex. road rage, person losing job & family becoming homicidal).

Frustration-aggression hypothesis According to this hypothesis, frustration occurs in situations in which people are prevented or blocked from attaining their goals; a rise in frustration then leads to a greater probability of aggression.

Full inclusion Policy or practice of placing all students in regular classes with appropriate assistance.

Full term babies Babies who arrive on or shortly before or after their due dates.

Function word A word such as *who* or *that* that makes explicit the relationship among various phrases within a sentence.

Functional amnesia A severe type of memory loss caused by psychological factors such as anxiety, hysteria, or repression.

Functional approach Attitude theories that emphasize that people develop and change their attitudes based on the degree to which they satisfy different psychological needs. To change an attitude, one must understand the underlying function that attitude serves.

Functional fixedness An inability to perceive a new use for an object previously associated with some other purpose; adversely affects problem solving and creativity.

Functional MRI (fMRI) A brain imaging technique that combines benefits of both MRI and PET scans by detecting magnetic changes in the flow of blood to cells in the brain.

Functionalism The perspective on mind and behaviour that focuses on the examination of their functions in an organism's interactions with the environment.

Fundamental attribution error (FAE) The dual tendency of observers to underestimate the impact of situational factors, and to overestimate the influence of dispositional factors on an actor's behaviour.

Fundamental frequency The lowest, and usually most intense, frequency of a complex sound; most often perceived as the sound's basic pitch.

G

G **factor** According to Spearman, a factor of intelligence that is common to all intellectual tasks; includes apprehension of experience, eduction of relations, and eduction of correlates.

GABA (Gamma Amino Butyric Acid) A neurotransmitter involved in the inhibition of anxiety and excitation. Too little GABA has been associated with anxiety disorders.

Galeophobia or Gatophobia Fear of cats.

Gallophobia or galiophobia Fear France or French culture. (Francophobia).

Galvanic skin response (GSR) A drop in the electrical resistance of the skin, widely used as an index of autonomic reaction.

Gambler's fallacy The false belief that a chance event is more likely if it hasn't happened recently.

Gamophobia Fear of marriage.

Ganglion cell A neuron in the retina that receives information from photoreceptors by means of bipolar cells, and from which axons proceed through the optic nerves to the brain.

Gate-control theory A theory about pain modulation that proposes that certain cells in the spinal cord act as gates to interrupt and block some pain signals while sending others on to the brain.

Gay bashing Attacks against homosexuals - either verbal or physical.

Gaze pattern In the study of channels of communication between individuals, the pattern of looking at the other that can have communicative value.

Geliophobia Fear of laughter.

Gender In psychology, the characteristics, whether biological or socially influenced, by which people define male and female. Because "sex" is a biological category, social psychologists sometimes refer to biologically based gender differences as "sex differences.".

Gender bias Different views of males and females, often favouring one gender over the other.

Gender constancy The recognition that being male or female is to all intents and purposes irrevocable.

Gender differences Culturally based differences between males and females.

Gender identity The internal sense of being either male or female. Usually congruent with biological gender, but not always as in Gender Identity Disorder.

Gender role Cultural expectations about the ways in which men and women should think and behave.

Gender schema theory Bem's theory that children develop schemas containing culturally based gender information which they use to understand themselves and the world.

Gender stereotypes A society's expectations about the characteristics of females as a group and males as a group.

Gender typing The process of developing the behaviours, thoughts, and emotions associated with a particular gender.

Gene The unit of hereditary transmission, located at a particular place in a given chromosome. Both members of each chromosome pair have corresponding locations at which there are genes that carry instructions about the same characteristic (e.g., eye colour). If one member of a gene pair is dominant and the other is recessive, the dominant gene will exert its effect regardless of what the recessive gene calls for. The characteristic called for by the recessive gene will only be expressed if the other member of the gene pair is also recessive.

General adaptation syndrome (GAS) The model proposed by Selye to describe the body's adaptation to chronic exposure to severe stressors. The body passes through an orderly sequence of three physiological stages: alarm, resistance, and exhaustion.

General factor Usually a factor on which all items have a positive factor loading, such as the factor known as general intelligence (g).

General intelligence The general intelligence factor (abbreviated g) is a widely accepted but controversial construct used in the field of psychology (see also psychometrics) to quantify what is common to the scores of all intelligence tests. The phrase "g theory" refers to hypotheses and results regarding g's biological nature, stability/malleability, relevance to real-world tasks, and other inquiries.

General intelligence factor (g) An ability that underlies all intelligent behaviour, proposed by Charles Spearman.

General paresis A psychosis characterized by progressive decline in cognitive and motor function culminating in death, reflecting a deteriorating brain condition produced by syphilitic infection.

Generalization In classical conditioning, CRs elicited by stimuli that resemble the CS used in training. In operant conditioning, the occurrence of responding when a stimulus similar (but not identical) to the discriminative stimulus is present.

Generalization decrement In classical conditioning, the weakening of a response to a new stimulus compared to the response elicited by the original conditioned stimulus (CS). The greater the difference between the new stimulus and the original CS, the larger the generalization decrement. In instrumental conditioning, a similar effect occurs when a new discriminative stimulus is presented instead of the original stimulus.

Generalization gradient The curve that shows the relationship between the tendency to respond to a new stimulus and its similarity to the original conditioned stimulus (CS).

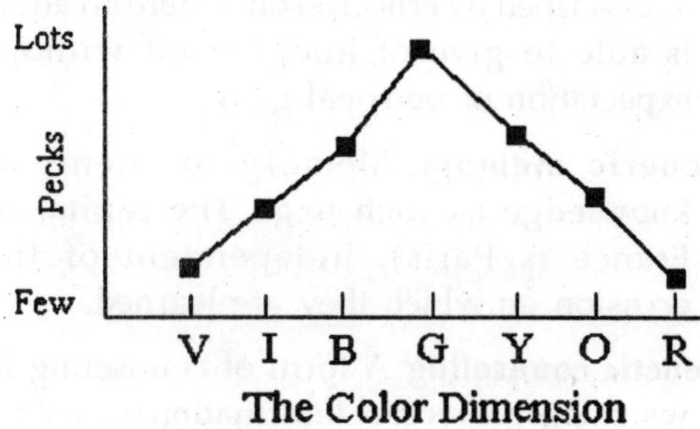

Fig. Generalization gradient

Generalize To formulate general principles or inferences from particulars.

Generalized anxiety disorder A mental disorder whose primary characteristic is an all-pervasive, "free-floating" anxiety. A member of the diagnostic category "anxiety disorders," which also includes phobias and obsessive-compulsive disorders.

Generalized cone The basic building block of marr's model of visual object representation, having a cross-section, an axis, and a description of the way that the cross-section changes along the axis.

Generalized other In self-concept formation, the global view we develop about how we are seen by others.

Generation-recognition A model of retrieval in which there is a generative phase and a recognition phase and which proposes that recall and recognition are different processes.

Generative The characteristic symbols of a language that can be combined to produce an infinite number of messages.

Generative grammar A way of thinking about grammar rules where they are seen as generating all and only those sentences that are legitimate in the language in question.

Generative learning A theory that emphasizes the active integration of new material with existing schemata.

Generativity One of the polarities of maturity, as described by erik erikson, where an adult is able to give of him/herself without expectation of personal gain.

Generic memory Memory for items of knowledge as such (e.g., The capital of France is Paris), independent of the occasion on which they are learned.

Genetic counseling A form of counseling in which people receive information regarding their family history of genetic disorders and the liklihood that they or their children may have a genetic disorder.

Genetic predisposition Those aspects of an individual that may lead them to respond more negatively to an unpleasant event, which are related to events they have experienced in the past.

Genetic sex A designation of an organism's sex based entirely on the genetic pattern, whether XY (male) or XX (female). Often contrasted with morphological sex, which is based on anatomical features.

Genetic variability The biological reality that offspring of any given species will differ from its parents, and that some offspring will survive by adapting to changing conditions.

Genetics The study of the genetic makeup of organisms and how it influences their physical and behavioural characteristics.

Geniophobia Fear of chins.

Genital stage In psychoanalytic theory, the stage of psychosexual development reached in adult sexuality in which sexual pleasure involves not only one's own gratification but also the social and bodily satisfaction brought to another person.

Genocide The systematic destruction of one group of people, often an ethnic or racial group, by another.

Genophobia Fear of sex.

Genotype The genetic blueprint of an organism which may or may not be overtly expressed by its phenotype.

Genuphobia Fear of knees.

Geon The basic building block of biederman's model of visual object representation, which is a simplified version of a generalized cone, having an axis and a cross-section which are defined only in terms of simple dichotomies (straight or curved, symmetrical or asymmetrical).

Gephyrophobia or gephydrophobia or gephysrophobia Fear of crossing bridges.

Gerascophobia Fear of growing old.

Germanophobia Fear of Germany or German culture.

Germinal stage The two-week period after conception.

Gerontologists Those who study the process of aging and of the problems of aged people.

Gerontology The scientific study of the process of aging and of the problems of aged people.

Gerontophobia Fear of old people or of growing old.

Gerstmann syndrome A syndrome that results from damage to the left parietal lobe which involves a cluster of difficulties, including the inability to form mathematical calculations, the loss of handwriting, and confusion about which direction is left and which is right.

Gestalt German word typically translated as meaning 'whole' or 'form.'.

Gestalt psychologists A group of mainly german psychologists, who were most active in the 1920s and interested, among other things, in the way that perception imposes structure upon its stimulus.

Gestalt psychology A school of psychology that maintains that psychological phenomena can be understood only when viewed as organized, structured wholes, not when broken down into primitive perceptual elements.

Gestalt therapy A form of therapy emphasizing the unity of mind and body by teaching the client to "get in touch" with unconscious bodily sensations and emotional feelings.

Gestation The act or period of carrying young in the uterus from conception to birth; pregnancy.

Geumaphobia or geumophobia Fear of taste.

Giftedness Category of exceptionality characterized by being very bright, creative, or talented.

Glands Bodily organs that produce hormones.

Glial cell Cells in the brain that act as guidewires for growing neurons, provide a supportive scaffolding for mature neurons, and form the myelin sheath and blood-brain barrier.

Glossophobia Fear of speaking in public or of trying to speak.

Glove anesthesia A condition sometimes seen in conversion disorders, in which there is an anesthesia of the entire hand with no loss of feeling above the wrist. This symptom makes no organic sense given the anatomical arrangement of the nerve trunks and indicates that the condition has a psychological basis.

Glucocorticoid A chemical, such as cortisol, that influences the metabolism of glucose, the main energy source of the body.

Glucose A simple sugar that acts as an energy source for cells.

Glucose receptors Receptors in the brain (in the area of the hypothalamus) that detect the amount of glucose in the bloodstream.

Glucostatic hypothesis The hypothesis that hunger is caused by a low level or availability of glucose, a condition that is monitored by specialized sensory neurons.

Glutamate The main excitatory neurotransmitter in the brain.

Glycogen A stored form of metabolic energy derived from glucose. To be used, it must first be converted back into glucose.

Gnosiophobia Fear of knowledge.

Goal structure The degree to which students are placed in competitive or cooperative relationships in earning classroom rewards.

Goal-directed selection A determinant of why people select some parts of sensory input for further processing; it reflects the choices made as a function of one's own goals.

God An idealized image of a nurturing (and primal) father created to reconcile us to Fate's cruelty, compensate us for the injustices of life, lend social moralities a divine origin, and personify and appease the uncontrollable forces of nature. He is a condensation of prior gods. His Judeo-Christian characteristics unconsciously permeate Freudian thought.

Gonads The body's primary sexual organs — ovaries in the female, testes in the male.

Good continuation A Gestalt law of organization; given two or more interpretations of elements that form the outline of the figure, the simplest interpretation will be preferred.

Good continutation A factor in visual grouping. Contours tend to be seen in such a way that their direction is altered as little as possible.

Good mood effect The effect whereby a good mood increases helping behaviour.

Grade equivalent scores Standard scores that relate students1 raw scores to the average scores obtained by norming groups a t different grade levels.

Graded receptor potential The response of a typical receptor cell involving a change in membrane potential that is proportional to the strength of the stimulus, rather than being an all-or-nothing action potential.

Gradualism In the study of evolution, the view that most of the time a species exists it will stay exactly the same, with only slow changes occurring.

Graduated and reciprocated initiatives in tension-reduction (GRIT) A strategy for unilateral, persistent efforts to establish trust and cooperation between opposing parties.

Grandiose delusion A belief centered around the idea that one is very important or famous.

Graphophobia Fear of writing or handwriting.

Gray matter That portion of the brain that appears gray. The colour reflects the absence of myelination (which makes the tissue appear white). The gray matter consists of the cell bodies, dendrites, and unmyelinated axons that comprise the nervous system's microcircuitry.

Great apes The larger group of apes - the orangutan, gorilla and two species of chimpanzee - which are the closest animal relatives to man.

Grief Intense emotional suffering caused by loss, disaster, misfortune, and so on; acute sorrow; deep sadness.

Grief therapy Treatment that helps individuals deal with their grief and bereavement.

GRIT Acronym for "graduated and reciprocated initiatives in tension reduction"- a strategy designed to de-escalate international tensions.

Gross motor skills The use of large bodily movements, including running, jumping, hopping, turning, skipping, throwing, balancing, and dancing.

Ground The backdrop or background areas of the visual field, against which figures stand out.

Group A collection of individuals who generally have common interests and goals.

Group alerting Methods of questioning that encourage students to pay attention during lectures and discussions.

Group cohesion In the study of groups, the development of friendships and the structuring of the group, which occurs as some individuals take the lead and others tend to follow.

Group cohesiveness The attractiveness that group members have for one another.

Group contingency program Program in which rewards or punishments are given to a class as a whole for adhering to or violating rules of conduct.

Group dynamics The study of how group processes change individual functioning.

Group factor theory An approach in the theory of intelligence, which emphasizes the independence of different factors that make up intelligence rather than one important general factor; using factor analysis, the American psychologist Thurstone identified seven separate factors in the performance of intelligence tests.

Group Investigating A cooperative learning model that involves small groups in which students work using cooperative inquiry, planning, project, and group discussion, then make a presentation on their findings to the class.

Group membership Acceptance of the individual into a particular social group, which in turn is a powerful influence in the formation of the self.

Group mind Notion that minds of persons are not totally separate and that minds of people belonging to a natural closely knit group merge to form some kind of collective consciousness.

Group polarization Group-produced enhancement of members' preexisting tendencies; a strengthening of the members' average tendency, not a split within the group.

Group psychology Freud distrusted groups, particularly large ones, almost as much as Jung did. Two key components of mass process: regressive identification with a powerful leader or idea, and replacement of one's ego ideal with that leader or idea.

Note: Freud wrote mass in German; this is usually mistranslated into "group.".

Group psychotherapy Therapy in which two or more clients meet simultaneously with a therapist, discussing problems within a supportive and understanding environment.

Group selection In sociobiology, the evolution of traits for the good of the species or group, even though they disadvantage the individual.

Group-serving bias Explaining away outgroup members' positive behaviours; also attributing negative behaviours to their dispositions (while excusing such behaviour by one's own group).

Group therapy Psychotherapy conducted with at least three or four non-related individuals who are similar in some are, such as gender, age, mental illness, or presenting problem.

Group think The tendency for members of a cohesive group to reach decisions without weighing all the facts, especially those contradicting the majority opinion.

Grouped frequency distribution A table showing the number of occurrences for a grouping of scores. Used a lot in educational settings where a score of 90 to 100 may be grouped as an A, a score of 80 to 90 may be grouped as a B, etc.

Groupthink A deterioration of mental efficiency, reality testing, and moral judgment in a group that results from an excessive desire to reach consensus.

Growth motives In self actualization, those motives which move a person up to the higher part of a hierarchy, to an achievement of things beyond the satisfaction of essential needs.

Growth need strength In the psychology of the work-place, a theory relating to the perception of job meaningfulness, where those low in growth need will not be motivated by increases in job characteristics - such as skill, responsibility, autonomy - whereas those high in growth need will benefit.

Growth needs Needs for knowing, appreciating, and understanding, which people try to satisfy after their basic needs are met.

Guevedoces syndrome A genetic disorder in which the external genitals appear to be female, but, at puberty, develop into normal male genitals.

Guided search In visual perception, a parallel search of the environment for single, basic attributes that guides attention to likely locations of objects with more complex combinations of attributes.

Guilt From either a dread of an external authority or dread of the demands and punishments of the superego, an internalized authority. It is through drive-repressing guilt and the resulting sublimations that civilization arose. Guilt also has a primeval source: the murder of the primal father .

Gustation Sense of taste.

Gymnophobia Fear of nudity.

Gynephobia or gynophobia Fear of women.

Gyri The ridges formed by adjacent clefts or sulci in the cerebrum.

H

Habituation A decline in the tendency to respond to stimuli that have become familiar. While short-term habituation dissipates in a matter of minutes, long-term habituation may persist for days or weeks.

Habituation procedure A widely used method for studying infant perception. After some exposure to a visual stimulus, an infant becomes habituated and stops looking at it. The extent to which a new stimulus leads to renewed interest and resumption of looking is taken as a measure of the extent to which the infant regards this new stimulus as different from the old one to which he became habituated.

Hadephobia Fear of hell.

Hagiophobia Fear of saints or holy things.

Hair cells The auditory receptors in the cochlea, lodged between the basilar membrane and other membranes above.

Hallucination False perception of reality (e.g., hearing voices that aren't there or seeing people who do not exist) [auditory (hearing); visual (sight); olfactory (smell); tactile (touch); and taste].

Halo effect The tendency to assign generally positive or generally negative traits to a person after observing one specific positive or negative trait, respectively.

Hamartophobia Fear of sinning.

Handedness Ability in using one hand more skillfully than, and in preference to, the other.

Handicap A condition imposed on a person with disabilities by society, the physical environment, or the person1s attitude.

Haphephobia or haptephobia Fear of being touched.

Haplorhines One of the two great branches of the primate order, comprised of monkeys and apes.

Hardiness Resistance to stress.

Hard-to-get effect The tendency to prefer people who are highly selective in their social choices over those who are more readily available.

Harmonic The simplest basic sound into which all auditory waveforms can be broken down, consisting of a sinusoid of a specific frequency and amplitude.

Harpaxophobia Fear of being robbed.

Hawthorne effect In the psychology of the workplace, the connection between increased productivity and the introduction of new working methods, the argument being that workers respond positively, if only temporarily, to any kind of innovation.

Haze A monocular cue of depth perception; objects that are less distinct in their outline and texture are seen as farther from the viewer.

Health A general condition of soundness and vigor of body and mind; not simply the absence of illness or injury.

Health promotion The development and implementation of general strategies and specific tactics to eliminate or reduce the risk that people will become ill.

Health psychology Health psychology is the 'use of psychological principles to promote health and to prevent illness' (Taylor, 1990). In other words, health psychology uses the knowledge and ideas from psychology to cut down on health risk factors, and promote healthy behaviours. Health risk factors include things like stress, overeating, and putting yourself in danger, including, for instance, not buckling one's seat belt, habits that increase chances of contracting illnesses. Being over-stressed is in fact a health risk factor, because stress can contribute to psychosomatic illnesses. Health promoting behaviours are any behaviour that promotes health, and include things like good exercise, little stress, buckling seat belts, wearing a helmet, and healthy eating. Using health psychology by cutting down on health risk factors and increasing healthy behaviours can greatly increase ones chances of living longer.

Hearing loss Degree of deafness; uncorrectable inability to hear well.

Hebb rule or Hebbian learning A natural form of learning which suggests that 'cells that fire together, wire together' - i.e. If two neurones are active at the same time and they synapse to each other, then the strength of the synapse will be increased.

Hedonophobia Fear of feeling pleasure.

Heliophobia Fear of the sun.

Hellenologophobia Fear of Greek terms or complex scientific terminology.

Helminthophobia Fear of being infested with worms.

Hemophobia or hemaphobia or hematophobia Fear of blood.

Heredity The sum of the traits and tendencies inherited from a person's parents and other biological ancestors.

Heresyphobia or hereiophobia Fear of challenges to official doctrine or of radical deviation.

Heritability Degree to which variation among individuals can be attributed to genetic factors.

Heritability estimate A statistical estimate of the degree of inheritance of a given trait or behaviour, assessed by the degree of similarity between individuals who vary in their extent of genetic similarity.

Heritability ratio (H) This refers to the relative importance of heredity and environment in determining the observed variation of a particular trait. More specifically, *H* is the proportion of the variance of the trait in a given population that is attributable to genetic factors.

Hermaphrodite A person who possesses portions of both female and male reproductive tissues.

Herpetophobia Fear of reptiles or creepy, crawly things.

Hertz (Hz) The primary measure of the frequency of vibration of sound waves; cycles per second.

Heteronomous morality Stage at which children think that rules are unchangeable and that breaking them leads automatically to punishment.

Heterophobia Fear of the opposite sex. (Sexophobia).

Heterosexism A system of cultural beliefs, values, and customs that exalts heterosexuality and denies, denigrates, and stigmatizes any nonheterosexual form of behaviour or identity.

Heterosexuality A primary or exclusive attraction to individuals of the other sex.

Heuristics A set of experimental, trial-and-error kinds of procedures used for assessing the probabilities of events or of properties that objects might have and for making inferences from those probabilities.

Hexakosioihexekontahexaphobia Fear of the number 666.

Hierarchical classification The ability to classify according to more than one level.

Hierarchical organization Organization in which narrower categories are subsumed under broader ones, which are subsumed under still broader ones, and so on. Often expressed in the form of a tree diagram.

Hierarchy of needs According to Maslow and other adherents of the humanistic approach, human needs are arranged in a hierarchy with physiological needs such as hunger at the bottom, safety needs further up, the need for attachment and love still higher, and the desire for esteem yet higher. At the very top of the hierarchy is the striving for self-actualization. By and large, people will only strive for the higher-order needs when the lower ones are fulfilled.

Hierophobia Fear of priests or sacred things.

High density setting In the evaluation of personal environment the level of proximity and crowding within buildings such as housing blocks or the locality in which people live, which may have an effect on their psychological well-being.

Higher-order conditioning In classical conditioning, a procedure by which a new stimulus comes to elicit the conditioned response (CR) by virtue of being paired with an effective conditioned stimulus (CS) (e.g., first pairings of tone and food, then pairings of bell and tone, until finally the bell elicits salivation by itself).

Higher-order patterns Patterns of ratios and relationships in the stimulus input.

Hindbrain Portion of the brain consisting of the medulla, the pons, and the cerebellum.

Hindsight bias The tendency to exaggerate, after learning an outcome, one's ability to have foreseen how something turned out. Also known as the i-knew-it-all-along phenomenon.

Hippocampus A major structure located within the forebrain and which forms part of the brain's limbic system, playing an important part in emotional behaviour and memory.

Hippophobia Fear of horses.

Hippopotomonstrosesquippedaliophobia Fear of long words.

Histogram A graphic rendering of a frequency distribution which depicts the distribution by a series of contiguous rectangles.

Historical research An examination of data to understand the past.

History External events that take place during a research study that are not part of the study but have an effect on the outcome.

Histrionic personality A personality type characterized by a desire to be the center of attention and the tendency to be self-focused, excitable, highly open to suggestion, very emotional, and dramatic.

Histrionic personality disorder A disorder characterized by attention-seeking behaviour and shallow emotions.

HIV Human immunodeficiency virus, a virus that attacks white blood cells (T lymphocytes) in human blood, thereby weakening the functioning of the immune system; HIV causes AIDS.

Hobophobia Fear of bums or beggars.

Hodophobia Fear of road travel.

Holophrase A single word that conveys complete ideas.

Holophrases Two-word utterances, such as 'allgone', 'oh-dear', which the child learns as a single word during early speech development.

Home range In environmental psychology, the particular area through which any animal species will move, on both a daily and seasonal basis, in order to gain the resources it needs.

Home-based reinforcement strategies Behaviour modification strategies in which a student1s school behaviour is reported to parents, who supply rewards.

Homeostasis The process by which important physiological characteristics (such as body temperature and blood pressure) are regulated so that they remain at their optimum level.

Homichlophobia Fear of fog.

Homilophobia Fear of sermons.

Hominophobia Fear of men.

Homogamy The tendency of like to mate with like.

Homophobia Fear of sameness, monotony or of homosexuality or of becoming homosexual.

Homosexual Of or characterized by sexual desire for those of the same sex as oneself.

Homosexuality A primary or exclusive attraction to individuals of one's own sex.

Homunculus A little person residing within the brain, from where he/she governs human behaviour (based on ancient Egyptian beliefs).

Hope That quality of ego functioning, as defined by Erik Erikson, which enables the infant to differentiate between trust and mistrust and develop a capacity for viewing life with optimism.

Hoplophobia Fear of firearms.

Horizontal cells The cells that integrate information across the retina; rather than sending signals toward the brain, horizontal cells connect receptors to each other.

Hormephobia Fear of shock.

Hormone A chemical released by one of the glands. Hormones travel through the bloodstream and control a number of bodily functions, including metabolic rate, arousal level, sugar output of the liver, and so on.

Hospice A homelike facility to provide supportive care for terminally ill patients.

Hostile aggression Aggression driven by anger and performed as an end in itself (also called affective aggression).

Hostile attributional bias A tendency for some people to perceive others' actions as the result of a hostile intent, when this is clearly not the case. Occurs in a number of criminal personality types.

Hostile witness This witness is deliberately antagonistic and/or noncompliant. Such an individual may invent facts in order to purposefully mislead the law. The hostile witness may react this way for two reasons:
1. their relationship to the offender, or

2. A underlying animosity towards the law.

Hozho A Navajo concept referring to harmony, peace of mind, goodness, ideal family relationships, beauty in arts and crafts, and health of body and spirit.

Hue A perceptual dimension of colour, most closely related to the wavelength of a pure light. The effect of a particular hue is caused by the mixture of lights of various wavelengths.

Human behaviour genetics The area of study that evaluates the genetic component of individual differences in behaviours and traits.

Human chorionic gonadotropin (HCG) The hormone that is secreted by the placenta early in pregnancy and which inhibits menstrual periods.

Humanism A school of thought that encourages seeing people's lives as those people would see them.

Humanistic approach to personality Asserts that what is most important about people is how they achieve their selfhood and actualize their potentialities.

Humanistic perspective A psychological model that emphasizes an individual's phenomenal world and inherent capacity for making rational choices and developing to maximum potential.

Humanistic psychology A theoretical view of human nature which stresses a positive view of human nature and the strong belief in psychological homeostasis.

Humanistic therapies Methods of treatment that emphasize personal growth and self-fulfillment. They try to be relatively nondirective, since their emphasis is on helping the clients achieve the capacity for making their own choices.

Humanistically oriented system of psychology Humanism is opposed to the mechanical view of man and humanistic psychology stresses the capacity of man for personal growth.

Human-potential movement The therapy movement that encompasses all those practices and methods that release the potential of the average human being for greater levels of performance and greater richness of experience.

Huntington's disease A progressive hereditary disorder that involves degeneration of the basal ganglia and that results in jerky limb movements, facial twitches, and uncontrolled writhing of the body.

Hyaluronidase An enzyme that inactivates hyaluronic acid by breaking down its polymeric structure, thus promoting the diffusion of substances through tissues found in sperm cells, certain venoms and bacteria, and so on.

Hydrargyophobia Fear of mercurial medicines.

Hydrophobia Fear of water or of rabies.

Hydrophobophobia Fear of rabies.

Hyelophobia or Hyalophobia Fear of glass.

Hygrophobia Fear of liquids, dampness, or moisture.

Hylephobia Fear of materialism or the fear of epilepsy.

Hylophobia Fear of forests.

Hypengyophobia or hypegiaphobia Fear of responsibility.

Hyperactivity Condition characterized by extreme restlessness and short attention spans relative to peers.

Hypercolumn The complete set of several hundred thousand cells in the primary visual cortex responsible for the preliminary analysis of one small region of the image.

Hypercomplex cell A type of cell in the visual cortex, also known as end-stopped cell, that is orientationally selective, having a relatively large receptive field, and which responds best to stimuli that end within its receptive field.

Hypermnesia The ability for the subject to recall otherwise forgotten memories. Typically, the subject will refer to the memories it in past, third-party tense. The opposite of Revivification.

Hyperphagia Voracious, chronic overeating brought about by a lesion of the ventromedial region of the hypothalamus.

Hypnagogic state An alternate state of consciousness at the onset of sleep, and the perceptions, fantasies, and energy levels provoked by that state.

Hypnobirth A method that teaches you that in the absence of fear and tension, pain does not necessarily have to be an accompaniment of labour.

Hypnoidal The earliest stage of hypnosis. Subject is drowsy, yet still aware. Arms and legs are typically warm and heavy.

Hypnophobia Fear of sleep or of being hypnotized.

Hypnosis (As defined by the American Psychological Association Division of Psychological Hypnosis) A procedure during which a health professional or researcher suggests that a client, patient, or subject experience changes in sensations, perceptions, thoughts, or behaviour. Any definition is necessarily vague, as the underlying mechanism is little understood. Some theories view hypnosis as an altered state of consciousness, others as a type of focused attention. Psychologists have recently researched hypnosis and found a strong correlation between the ease of putting someone in a state of 'hypnosis' and their level of suggestibility.

Hypnotherapy The application of hypnosis as a form of medical treatment, particularly for conditions largely related to the person's state of mind. Practitioners believe that when a client enters a trance state, or believes to do so, the client's resistance to beneficial change will be lowered significantly, and so various means of therapy will be easier to deliver and more successful.

Hypnotist An individual, most likely unlicensed, who uses hypnosis techniques or variations of these techniques for a variety of reasons, including treatment and/or entertainment.

Hypnotizability The degree to which an individual is responsive to standardized hypnotic suggestion.

Hypochondriasis A somatoform disorder involving persistent and excessive worry about developing a serious illness. People with this disorder often misinterpret the appearance of normal physical aches and pains.

Hypomania A mild manic state in which the individual seems infectiously merry, extremely talkative, charming, and indefatigable.

Hypothalamus A small structure located inside the brain which is important in numerous motivational processes such as hunger, thirst and sex and which also plays a role in controlling emotion, sleep and bodily temperature. In addition, it can exert certain physical and emotional effects by directing the release of hormones from the pituitary gland.

Hypothesis An unproved theory, proposition, supposition, and so on, tentatively accepted to explain certain facts or to provide a basis for further investigation, argument, and so on.

Hypothesis testing A procedure for testing the truth or usefulness of a hypothesis, by deducing specific consequences from it and finding out whether those consequences are true.

Hypsiphobia Fear of height.

Hysterectomy The surgical removal of all or part of the uterus.

Hysteria An older term for a group of presumably psychogenic disorders including conversion disorders and dissociative disorders. Since DSM-III, it is no longer used as a diagnostic category, in part because of an erroneous implication that the condition is more prevalent in women (Greek *hystera* – womb).

I

Iatrophobia Fear of going to the doctor or of doctors.

Ichthyophobia Fear of fish.

Iconic memory A form of sensory memory that holds a brief visual image of a scene that has just been perceived.

Id (Es) The permanently unconscious motivational cauldron of the mind. From the id (the "it") originate all the drives that impel psychic life. A "residue of countless egos" inherited from prior generations, the id is the amoral beast within us that seeks only its own gratification through tension discharge. It is powered by the bodily instincts and is wholly irrational. Analogous to the job of the imperialist and the industrialist, the job of the ego is to dominate it. (The term id comes from Groddeck, who got it from Nietzsche.)

Note: Freud wrote "the it" in German; this use should be retained rather than the latinized id.

Ideal self Humanistic term representing the characteristics, behaviours, emotions, and thoughts to which a person aspires.

Ideas of reference A characteristic of some mental disorders, notably schizophrenia, in which the patient begins to think that external events are specially related to her personally (e.g., "People walk by and follow me").

Identical triplets Three foetuses produced from the same ovum.

Identical twins Twins that originate from a single fertilized egg that then splits into two exact replicas that develop into two genetically identical individuals.

Identification An early, primitive kind of attachment to an object which results in incorporating some of its aspects into oneself. Ego and superego make use of identification to attract libido away from objects and toward themselves, thereby building up the personality. Other types include narcissistic, goal-oriented, object-loss, and aggressor identification.

Identification and recognition Two ways of attaching meaning to percepts.

Identity The condition or fact of being a specific person or thing; individuality.

Identity achievement A state in which a person commits to an identity after considering alternative possibilities.

Identity crisis The condition of being uncertain of one's feelings about oneself, especially with regard to character, goals, and origins, occurring especially in adolescence as a result of growing up under disruptive, fast-changing conditions.

Identity diffusion A state of confusion when a person lacks a clear sense of identity and hasn't yet begun exploring issues related to identity development.

Identity foreclosure A state in which a person has prematurely committed to values or roles prescribed by others.

Identity moratorium A state in which commitment to an identity is delayed while a person experiments with various roles and values.

Identity status Four categories of reaction in young people to certain questions posed them in a semi-structured interview and which have been defined as 'confusion', 'foreclosure', 'moratorium' and 'achievement'.

Ideology A set of beliefs and values held by the members of a social group, which explains its culture both to itself and to other groups.

Ideophobia Fear of ideas.

Idiosyncrasy credit In the study of social influence processes, a pre-existing, powerful role displayed by a 'minority' member of a particular group which can lead the group members to comply unexpectedly with them.

Idiot savant A child or adult who despite often (but not always) having a mental handicap has an extraordinary ability to perform mental calculations or demonstrate feats of knowledge, in many cases relating to numbers.

If-then profile In trait research, a formulation for analysing data types of behaviour by plotting the conditional if...then's of the particular responses to particular social situations.

Illocution In speech act theory, the intention of the speaker that is actually transmitted to the listener, which can be very different from the locution itself.

Illusion An experience of a stimulus pattern in a manner that is demonstrably incorrect but shared by others in the same perceptual environment.

Illusion of control Perception of uncontrollable events as subject to one's control or as more controllable than they are.

Illusion of out-group homogeneity A belief that members of groups to which one does not belong are very similar to one another.

Illusory conjunction A pattern of errors found, for example, in visual search tasks, in which observers correctly perceive the features present (redness, greenness, roundness, angularity) but misperceive how these were combined in the display (and so they might see a green O and a red X when, in fact, a green X and red O were presented).

Illusory contours Contours perceived in a figure when no contours are physically present.

Illusory correlation 1. perception of a relationship where none exists, or perception of a stronger relationship than actually exists.

2. A false impression that two variables correlate.

Illyngophobia Fear of vertigo or feeling dizzy when looking down.

Imageability In environmental psychology, the symbolic significance of place related to its distinctiveness of form, its visibility, and patterns of use.

Imagery Use of mental images to improve memory.

Immune system A network of organs and cells that protects the body from invading bacteria, viruses, and other foreign substances.

Immunoglobulins The antibodies that are released by B lymphocytes.

Immunological functioning In the study of psychological stress, the body's ability to resist infection and illness, a state which can be affected to low moods.

Implantation The process by which the embryo becomes embedded in the wall of the uterus.

Implicit attitudes Beliefs that are unconscious but that can still influence decisions and behaviour.

Implicit Attitudes Test (IAT) A method of measuring biological and psychological reaction time when a pair of values are presented.

Implicit egotism A nonconscious form of self-enhancement.

Implicit memory Any memory task which tests memory indirectly, e.g. Repetition priming.

Implicit personality theory A network of assumptions people make about the relationships among traits and behaviours.

Implicit uses of memory Availability of information through memory processes without the exertion of any conscious effort to encode or recover information.

Implosion therapy A form of behaviour therapy related to flooding in which the patient exposes himself to whatever he is áfraid of in its most extreme form, but does so in imagination rather than in real life (e.g., a person afraid of dogs has to imagine himself surrounded by a dozen snarling Dobermans).

Impossible figure A figure that appears acceptable when looked at locally but poses unresolvable visual contradictions when seen as a whole.

Impression formation The process by which one integrates various sources of information about another into an overall judgment.

Impression formation The way in which we form impressions of others and attribute specific characteristics and traits to them.

Impression management According to Erving Goffman, the characteristic of much social interaction in which people maintain the image that goes along with their social or professional role.

Impression regulation In the dyad, the control over the output from oneself to the other.

Imprinting A learned attachment that is formed at a particular period in life (the critical, or sensitive, period) and is difficult to reverse (e.g., the duckling's acquired tendency to follow whatever moving stimulus it encounters twelve to twenty-four hours after hatching).

Impulsive aggression Emotion-driven aggression produced in reaction to situations in the "heat of the moment.".

Impulsivity Cognitive style of responding quickly but often without regard for accuracy.

Inappropriate affect Expressing contradictory behaviour when describing or experiencing an emotion (e.g., smiling when discussing something sad; laughing when talking about the death of a loved one).

Incentive An environmental stimulus that pulls people to act in a particular way.

Incest The mating of close relatives who share many of the same genes.

Incidental learning Learning without trying to learn (e.g., as in a study in which participants judge a speaker's vocal quality when she recites a list of words and are later asked to produce as many of the words as they can recall).

Inclusive fitness In the study of evolution, the theory that inherited traits take account of all effects on relatives as well as the self, being the sum of their effects on the reproduction of the individual itself, plus the effect on the reproduction of all of its relatives, each devalued by the appropriate coefficient of relatedness.

Incongruence According to Carl Rogers and other humanistic therapists, a disparity between the self-concept and reality.

Incremental validity The extent to which a test adds to the predictive validity already provided by other measures (e.g., the extent to which a projective technique adds to what is already known through an ordinary interview).

Incubation The hypothetical process of continuing to work on a problem unconsciously after one has ceased to work on that problem consciously. Most contemporary investigators are skeptical about whether such a process truly exists.

Independence Not being subject to control by others.

Independent construals of self Conceptualization of the self as an individual whose behaviour is organized primarily by reference to one's own thoughts, feelings, and actions, rather than by reference to the thoughts, feelings, and actions of others.

Independent practice Component of instruction in which students work by themselves to demonstrate and rehearse new knowledge.

Independent samples Sample data that is independent or not related to each other.

Independent variable In experimental settings, the stimulus condition whose values are free to vary independently of any other variable in the situation.

Indifferent parents Parents who demonstrate low parental control and low warmth.

Indigenous healing Non-Western, culture-bound approaches to the treatment of psychological and medical problems.

Indirect perception The approach to perception that emphasizes the indirectness of the relationship between the final conscious impression and the initial pattern of stimulation and which suggests that perception is an active process involving sophisticated knowledge.

Individual Learning Expectation (ILE) A teaching method that includes evaluation of students improvement relative to past achievement.

Individual psychology Alfred Adler's school of thought, which maintains that the main motivations for human behaviour are not sexual or aggressive urges but strivings for superiority.

Individualism A cultural pattern in which people are considered to be fundamentally independent and in which the emphasis is on the ways a person can stand out through achieving private goals. Individualists societies include the dominant cultures of the United States, western Europe, Canada, and Australia.

Individualized Education Program (IEP) Program tailored to the needs of an exceptional child.

Individualized instruction Teaching approach in which each student works at his or her own level and rate.

Individuation The process, generally experienced in the transitional period between early and middle adulthood, whereby an adult becomes more reflective about his/her life and increasingly compassionate, accepting and more loving of self and others.

Induced motion An illusion in which a stationary point of light within a moving reference frame is seen as moving and the reference frame is perceived as stationary.

Induction A non-deductive type of reasoning where, even if a person knows the premises given are true, and knows that they have reasoned as they intended, they cannot be sure that their conclusion will be true, the end result being a hypothesis.

Inductive reasoning A form of reasoning in which a conclusion is made about the probability of some state of affairs, based on the available evidence and past experience.

Industrial/organizational psychology The area or specialty in psychology focused on the application of psychological principles in the work force.

Industry The feeling of social competence; according to Erikson, the primary developmental task of middle childhood is to attain industry.

Inert knowledge Learned information that can be applied to only a restricted, often artificial set of circumstances.

Infancy Birth to age 1.

Infant mortality The percentage of babies who die within the first year of life.

Infatuatio Completely carried away by foolish or shallow love or affection.

Inferences Missing information filled in on the basis of a sample of evidence or on the basis of prior beliefs and theories.

Inferential explosion The point arrived at when the use of rules to reason about a representation become vulnerable and therefore unsustainable.

Inferential statistics The branch of statistics that focuses on describing in numerical format what might be happening or what might happen (estimation) in the future (probability). Inferential statistics required the testing of only a sample of the population. (Example: 100 students rather than all students).

Inferiority complex An exaggerated sense of inferiority.

Inferred reality The meaning of stimuli in the context of relevant information.

Infertile The inability to produce offspring because of some disorder of the reproductive system.

Inflection A change in the form of a word (usually by adding a suffix) to denote a grammatical feature such as tense or number.

Information campaigns Attempts to persuade people to alter their lifestyles in more healthful directions through the use of the mass media and other communication channels.

Information dependence Dependence upon others for information about the world that reduces uncertainty.

Information integration theory The theory that impressions are based on 1. perceiver dispositions and

2. A weighted average of a target person's traits.

Information processing A general term for the presumed operations whereby the crude raw materials provided by the senses are refashioned into items of knowledge. Among these operations are perceptual organization, comparison with items stored in memory, and so on.

Information processing An approach used by cognitive psychologists to explain the workings of the brain; information received through the senses is processed by systems of neurons in the brain.

Informational influence Group effects that arise from individuals' desire to be correct and right and to understand how best to act in a given situation.

Informational social influence An individual's tendency to conform because a group provides one with information.

Information-processing theory Cognitive theory of learning that describes the processing, storage, and retrieval of knowledge from the mind.

Informed consent A procedure by which people freely choose to participate in a study only after they are told about the activities they will perform.

Infradian rhythms Biological cycles that take longer than twenty-four hours.

Ingratiation A technique used to obtain compliance by inducing someone to like you, then attempt to change their behaviours (i.e. cult behaviour, could be used with 'good cop' interrogations).

Ingroup A group to which a person belongs and that forms a part of his or her social identity.

Ingroup bias The tendency to give more favourable evaluations and greater rewards to ingroup members than to outgroup members.

Ingroup favouritism favouritism The tendency to discriminate in favour of ingroups over outgroups.

Inhalant Substances such as spray paint, freon, and glue that produce an intoxicating effect when inhaled.

Inherited behavioural differences Human characteristics, such as shyness, which may result from inherited genetic traits.

Inhibition In social psychology, the restraining effect on the individual of the presence of others when they are performing certain tasks, which stems from their apprehension at being evaluated by the other.

Inhibitory inputs Information entering a neuron signaling it not to fire.

Inhibitory postsynaptic potential A negative change in voltage that occurs when a neurotransmitter binds to an inhibitory receptor.

Initial-letter strategy Strategy for memorization in which initial letters of a list to be memorized are taken to make a word or phrase that is more easily remembered.

Initiative The ability to think and act without being urged; enterprise.

Injunctions Messages received during childhood.

Innate abilities Abilities that are present from birth.

Inner ear The portion of the ear in which the actual transduction of sound takes place.

Inoculation hypothesis The idea that exposure to weak versions of a persuasive argument increases later resistance to that argument.

Inquisitorial model A dispute-resolution system in which a neutral investigator gathers evidence from both sides and presents the findings in court.

Insanity The legal (not clinical) designation for the state of an individual judged to be legally irresponsible or incompetent.

Insectophobia Fear of insects.

Insecure attachment The lack of attachment by an infant to a particular care giver, so that it does not use the attachment figure as a secure base and when distressed will show some ambivalence to or avoidance of that attachment figure.

Insertional necrophilia As a substitute of sexual intercourse the offender inserts foreign objects into the victims orifice. Common with disorganized offenders. This should not be mistaken as a form of mutilation.

Insight The understanding of a relationship between current thoughts, feelings, and/or behaviours and where these originated or how they are maintained.

Instinct A behaviour we are born with and therefore does not need to be learned.

Insight therapies Treatments involving complex conversations between therapists and clients. The treatments aim to help clients understand the nature of their problems and the meaning of their behaviour, thoughts, and feelings.

Insightful learning Learning by understanding the relations among the components of the problem; often contrasted with "blind trial and error" and documented by wide and appropriate transfer if tested in a new situation.

Insomnia A form of sleep disorder characterised by an inability to sleep and/or to remain asleep for a reasonable period during the night. Sufferers typically complain of being unable to close their eyes for more than a few minutes at a time, or of 'tossing and turning' through the night.

Instinct Freud discussed instincts, which are relatively unchangeable, primarily in connection with animal life, not human life.

Instinct theory Aggression theory that suggests aggression stems from universal innate tendencies. In other words aggression is born into all of us.

Instinctive behaviour An innate, unlearned behaviour pattern exhibited by all members of a species.

Instinctive drift The tendency for conditioning to be hindered by natural instincts.

Instinctual drift The tendency for learned behaviour to drift toward instinctual behaviour over time.

Instinctual mechanisms Principles of expression and operation of instincts.

Instructional objective A statement of information or tasks that students should master after one or more lessons.

Instrumental aggression Aggression with the goal not to harm but rather to attain some other goal (i.e. sport aggression).

Instrumental conditioning A form of learning in which a behaviour becomes more or less probable, depending on it consequences. Rewards increase the probability that the behaviour will be repeated, whereas punishment or nc reward reduces the probability.

Instrumental Enrichment A thinking-skills program in which students work through a series of paper-and-pencil exercises designed to develop various intellectual abilities.

Instrumental learning Learning by drive reduction, law of effect or reinforcement. This type of learning presupposes a high drive to solve a problem, a reward or punishment and selective strenthening or weakening of connections.

Instrumentation The reliability of the instrument may change in calibration (if using a measuring device) or from change in human ability to measure differences (due to fatigue, experience, etc). A threat to internal validity.

Insufficient deterrence A condition in which people refrain from engaging in a desirable activity, even when only mild punishment is threatened.

Insufficient justification A condition in which people freely perform an attitude-discrepant behaviour without receiving a large reward.

Insufficient justification effect Reduction of dissonance by internally justifying one's behaviour when external justification is "insufficient.".

Insulin A hormone with a crucial role in using nutrients. One of its functions is to help promote the conversion of glucose into glycogen.

Integrated internal representation The method applied by a blind child for building up an awareness of a particular area, which is relative to the amount of lifetime freedom it has been allowed to explore space by its parents or carers.

Integrated learning system A comprehensive, multipurpose set of instructional software developed by one company.

Integrative agreement A negotiated resolution to a conflict in which all parties obtain outcomes that are superior to what they would have obtained from an equal division of the contested resources.

Integrative approach Therapy approaches that combine the ideas and techniques of several different schools of psychology.

Integrity test A paper-and-pencil questionnaire designed to test a job applicant's honesty and character.

Intelligence Intelligence is usually said to involve mental capabilities such as the ability to reason, plan, solve problems, think abstractly, comprehend ideas and language, and learn. Although laymen generally regard the concept of intelligence as having much broader scope, in psychology, the study of intelligence generally regards this personality trait as distinct from creativity, personality, character, or wisdom.

Intelligence quotient (IQ) A number intended to indicate a person's level of intelligence. It is the mental age (as shown by intelligence tests) multiplied by 100 and divided by the chronological age.

Intelligence scale A scale devised by binet and simon, which was based on already-existing tests of memory, comprehension, vocabulary and reasoning, the selection being based on items which generated good performance from children rated as bright as well as poor performance from children considered less able.

Intensity cue One of the main sources of information in locating the direction of sound sources. Sounds coming from the left, for example, will be more intense in the left ear than in the right.

Intentional deception One of the thorniest of ethical issues in psychological research, deception is used in psychological studies through the use of confederates in order to gauge the response of subjects to wrong answers. The use of deception is usually revealed to the subject in the debriefing session.

Intentional learning Learning when informed that there will later be a test of learning.

Inter quartile range The middle 50% of the scores, as determined by the median. A good measure of variability if there are many outliers.

Interaction A statistical term indicating the change in the effect of each independent variable as a function of other independent variables.

Interaction effect What is seen when different levels of the IV and DV interact to give different results for different groups.

Interactional synchronicity A stage of development whereby babies over the age of 2 or 3 months develop a readiness for communicative interaction with their carers.

Interactional theory of homosexuality A theory stating that sexual orientation develops from a complex interaction of biological, psychological, and social factors.

Interactionalist resolution In the analysis of personality dimension, the study of behaviour as a product of traits expressed in interaction with social situations.

Interactionism An important perspective in social psychology that emphasizes the combined effects of both the person and the situation on human behaviour.

Interactionist perspective An emphasis on how both an individual's personality and environmental characteristics influence behaviour.

Interaural delay A difference in the time of arrival of a stimulus at the two ears. An important source of information about the direction of a sound source.

Interdependent construals of self Conceptualization of the self as part of an encompassing social relationship; recognizing that one's behaviour is determined, contingent on, and, to a large extent organized by what the actor perceives to be the thoughts, feelings, and actions of others.

Interference A memory phenomenon that occurs when retrieval cues do not point effectively to one specific memory.

Interference theory of forgetting The hypothesis that items are forgotten because they are somehow interfered with by other items learned before or after.

Intergroup Awareness and hostility in social psychology, the existence of prejudice and hostility as part of the normal relations within groups.

Intergroup comparison In social identity theory, the comparison between one's own group and other groups.

Intermale aggression Physical violence or submissive behaviour displayed by males towards each other.

Intermittent reinforcement A reinforcement schedule in which reinforcement happens only on some of the occasions a particular response occurs. It is also called *partial reinforcement*.

Internal attribution An attribution that locates the cause of an event to factors internal to the person, such as personality traits, moods, attitudes, abilities, or effort.

Internal consistency A measure of reliability; the degree to which a test yields similar scores across its different parts, such as on odd versus even items.

Internal locus of control The belief that an individual has more control over life circumstances than the environment does.

Internal validity A measure of the trustworthiness of a sample of data. Internal validity looks at the subject, testing, and environment in which the data collection took place.

Internalization The process whereby moral codes are adopted by the child so that they control her behaviour even when there are no external rewards or punishments.

Interneurons Brain neurons that relay messages from sensory neurons to other interneurons or to motor neurons.

Interobserver reliability There are consistent results among different testers or coders who are rating the same information.

Interpersonal attraction A person's desire to approach another individual.

Interposition A monocular depth cue in which objects that are farther away are blocked from view by any other opaque object obstructing their optical path to the eye.

Interpretation A psychoanalytic technique that involves suggesting the hidden meanings of free associations, dreams, feelings, memories, and behaviour to the client.

Interquartile range The difference between the scores (or estimated scores) at the 75th percentile and the 25th percentile. Used more than the range because it eliminates extreme scores.

Interrater reliability The degree to which different observers agree on their observations.

Intersexual A child who is not clearly male or female, in some cases because of genetic factors, in others because of morphology.

Interval estimation Estimating the population statistic based on a range around a sample statistic.

Interval measurement uses a set scale: the researcher knows the order of the data points both relative to each other and in absolute terms. The zero has no meaning, but there is a meaningful average.

Interval scale A scale in which equal differences between scores can be treated as equal so that the scores can be added or subtracted.

Interval schedule A reinforcement schedule in which reinforcement is delivered for a first response made after a given interval of time has elapsed. In a fixed-interval schedule, the interval is always the same. In a variable-interval schedule, the interval varies around a specified average.

Interview A subjective personality and mental health assessment typically consisting of questions and answers.

Interviewer bias An error in research that occurs when an interviewer's expectations or gestures influence a subject's responses.

Intimacy The sense of warmth and closeness in a loving relationship, including the desire to help the partner, self-disclose, and keep him or her in one's life.

Intimate relationship A close relationship between two adults involving at least one of the following: emotional attachment, fulfillment of psychological needs, and interdependence.

Intimidated witness This witness fears retaliation from the offender(s) involved or from the criminal element in general. As a result, the intimidated witness is quite apprehensive.

Intimidation Arousing fear and gaining power by convincing others that one is dangerous.

Intracranial recording The recording of brain activity by monitoring chemical or electrical activity from inside the skull.

Intraspecific aggression The attack by one animal upon another member of its species.

Intrinsic incentive An aspect of an activity that people enjoy and, therefore, find motivating.

Intrinsic motivation The motivation or desire to do something based on the enjoyment of the behaviour itself rather than relying on or requiring external reinforcement.

Intrinsic reinforcer The pleasure that is inherent in simply engaging in the behaviour.

Introjection Ferenczi's term for the psychological action by which a person is internalized and made a part of one's own psyche.

Introspection The most informal kind of self-report, in which a subject speculates on his/her cognitive states and processes, and which produces data of a highly qualitative nature.

Introversion Jung's term for a person whose mind is internally turned. In modern Psychology, a person who is withdrawn, restful and prefers to be alone.

Invariant A feature of the stimulus that reliably signals some useful feature of the world and which is usually associated with the direct approach to perception.

Inventive witness These witnesses typically embellish and create details when being interviewed. This is likely due to an inability to differentiate between fantasy and reality. They may also have esteem motives, in that they wish to appear important.

Inverse relationship High values on one variable tend to occur with low values on the other variable.

Inversion 1. Homosexuality, the cause of which, for Freud, was in the boy's failure to disidentify with Mom and identify with Dad and in the girl's identification with Dad rather than Mom.

2. The dream's transformation of one thing into its opposite in order to disguise it.

Ion A positively or negatively charged particle; produced when many substances dissolve in water.

Ion channel A special protein molecule located in the membrane of a cell; controls the entry or exit of particular ions.

Ion transporter A special protein molecule located in the membrane of a cell; actively transports ions into or out of the cell.

Iophobia Fear of poison.

Iris The smooth circular muscle in the eye that surrounds the pupil and contracts or dilates under reflex control in order to govern the amount of light entering.

Irreversibility The inability to mentally reverse an operation.

Irritable aggression Aggression and rage directed towards an object when the aggressor is frustrated, hurt, deprived, or stressed. As a result one may aggress towards objects as an acceptable outlet of the aggression.

Isolation A mechanism of defence in which anxiety arousing memories are retained but without the emotion that accompanied them.

Isolation aphasia A language disturbance that includes an inability to comprehend speech or to produce meaningful speech without affecting the ability to repeat speech and to learn new sequences of words; caused by damage that isolates the brain's speech mechanisms from other parts of the brain.

Isolophobia Fear of solitude, being alone.

Isopterophobia Fear of termites, insects that eat wood.

Ithyphallophobia Fear of seeing, thinking about or having an erect penis.

J

James Lange theory of emotion A peripheral-feedback theory of emotion stating that an eliciting stimulus triggers a behavioural response that sends different sensory and motor feedback to the brain and creates the feeling of a specific emotion.

Japanophobia Fear of Japanese.

Jargon The second stage of speech development in the child after babble, where the child forms syllable sequences which increasingly use the sound com-binations, rhythms and intonation patterns of the language it hears.

Jaspers, Karl Inspired by Nietzsche's and Kierkegaard's psychology, but aiming at a strictly scientific method, the "existen-tialist" Jaspers analyzes the possible attitudes of man towards the world; the decisions which the individual must make in inescapable situations like death, struggle, change, guilt; and the various ways in which man meets these situations. Motivated by the boundless desire for clarity and precision, Jaspers earnestly presents as his main objective to awaken the desire for a fuller, more genuine philosophy, these three methods of philosophizing which have existed from te earliest times to the present: Philosophical world orientation consisting in an analysis of the limitations, incompleteness and relativity of the researches, methods, world pictures of all the sciences; elucidation of existence consisting of a cognitive penetration into reality on the basis of the deepest inner decisions experienced by the individual, and striving to satisfy the deepest demands of human nature; the way of metaphysics, the never-satisfied and unending search for truth in the world of knowledge, conduct of life and in the seeking for the one being, dimly seen through antithetic thoughts, deep existential conflicts and differently conceived metaphysical symbols of the past. Realizing the decisive problematic relation between philosophy and religion in the Middle Ages, Jaspers elevates psychology and history to a more important place in the future of philosophy.

Jealousy The negative emotional reaction experienced when a relationship that is important to a person's self-concept is threatened by a real or imagined rival.

Jigsaw classrooms Classrooms that use a technique known as jigsawing, in which each pupil is given part of the total material to master and then share with other group members.

Jnana yoga Reaching the absolute through the intellect, reflecting on the ultimate questions of existence.

Job burnout The syndrome of emotional exhaustion, depersonalization, and reduced personal accomplishment, often experienced by workers in high-stress jobs.

Joint attention The ability of a baby after the age of approximately 5 months to look where another person is looking.

Joplin Plan A regrouping method in which students are assigned to groups for reading instruction across grade lines.

Judeophobia Fear of Jews.

Judgment The process by which people form opinions, reach conclusions, and make critical evaluations of events and people based on available material; also, the product of that mental activity.

Jumping genes The theory that genes change position influencing the expression of characteristics.

Jung, Carl A student of Freud who split from the Psychoanalytic Society because of his disagreements with Freud, especially his view of the collective unconscious.

Jury nullification The jury's power to disregard, or "nullify," the law when it conflicts with personal conceptions of justice.

Just noticeable difference In psychophysics, the just noticeable difference (usually abbreviated as jnd, using lowercase letters) is the smallest difference in a sensory input that is perceivable by a human being or other animal. It is also known as the differential threshold.

Just world hypothesis The tendency to believe that the world is fair and that.

Justification of effort The idea that if one works hard to reach a goal, one is likely to value that goal.

Juvenile delinquency The breaking of the law by minors.

K

Kainolophobia or Kainophobia Fear of anything new, novelty.

Kakorrhaphiophobia Fear of failure or defeat.

Karma yoga The yoga of purifying consciousness through detached action (service).

Katagelophobia Fear of ridicule.

Kathisophobia Fear of sitting down.

Keller Plan A form of mastery learning in which students advance through the curriculum by passing mastery tests.

Kenophobia Fear of voids or empty spaces.

Keraunophobia or ceraunophobia Fear of thunder and lightning.(Astraphobia, Astrapophobia).

Keyword method Strategy for improving memory by using images to link pairs of items.

Kin selection Kin selection was first suggested by Darwin as an explanation of the sterile castes of social insects and has later been mathematically defined by W. D. Hamilton as a mechanism for the evolution of apparently altruistic acts. Kin altruism is the technical term for altruistic behaviour that has been shown, or is theoretically supposed, to be explained by kin selection.

Kinesthesis The sense of the position and movement of body parts.

Kinesthetic sense Sense concerned with bodily position and movement of the body parts relative to each other.

Kinetophobia or kinesophobia Fear of movement or motion.

Kleptophobia Fear of stealing.

Koinoniphobia Fear of rooms.

Kolpophobia Fear of genitals, particularly female.

Koniophobia Fear of dust. (Amathophobia).

Kopophobia Fear of fatigue.

Kosmikophobia Fear of cosmic phenomenon.

Kurtosis The shape of a curve or distribution of scores.

Kymophobia Fear of waves. (Cymophobia).

Kynophobia Fear of rabies.

Kyphophobia Fear of stooping.

L

Labor The process or period of childbirth; parturition; especially, the muscular contractions of giving birth.

Laboratory experiment An experiment that, unlike a field experiment is strictly controlled and far from natural. It can however make up for taking the phenomenon out of its context by allowing much more careful control of the different variables involved.

Laboratory observaticn An observational research method in which information about subjects is collected in a laboratory setting.

Lachanophobia Fear of vegetables.

Laissez faire Absence of proper leadership or leadership which lets people do what they want to do.

Laliophobia or lalophobia Fear of speaking.

Lamarckian theory The theory that acquired or learned characteristics is genetically transmitted to the offsprings. Though the mechanism of how learning can influence genes is not known there is some evidence that experience is genetically transmitted.

Lamaze method A training program in natural childbirth, emphasizing breathing control and relaxation during labor together with the presence and encouraging assistance of a partner.

Language A system of symbols and rules used for meaningful communication.

Language acquisition device (LAD) Proposed biologically based mental structure that theorists believe plays a major role in children's language learning. Linguist Noam Chomsky revolutionized the idea that an infant's innate ability to understand a language structurally, before actually being able to speak it, allows for the possibility that children can learn any language intuitively before a certain age.

Language acquisition support system (LASS) The conclusion of american psychologist jerome bruner that for a child to learn language normally it must have a support system of social interaction based on highly familiar play or care routines which support each next step of its language development.

Language disorders Impairments in the ability to understand language or to express ideas in one1s native language.

Language-making capacity The innate guidelines or operating principles that children bring to the task of learning a language.

Language minority Term for native speakers of any language other than English.

Language production What people say, sign, and write, as well as the processes they go through to produce these messages.

Language universal A characteristic feature found in all natural languages.

Large muscle development Movements, such as running or throwing, that involve the limbs and large muscles.

Late adulthood Those years that encompass age 65 and beyond, according to Daniel Levinson.

Late phase of labor The phase of labor in which contractions become very painful and the cervix dilates completely to 10 cm, or 4 inches.

Latency General term for the interval before some reaction occurs. Often *response latency*.

Latency period In psychoanalytic theory, a stage in psychosexual development in which sexuality lies essentially dormant, roughly from ages five to twelve.

Latency stage Freud's fourth stage of psychosexual development where sexuality is repressed in the unconscious and children focus on identifying with their same sex parent and interact with same sex peers.

Latent content In Freudian dream analysis, the hidden meaning of a dream.

Latent content The true thoughts below the manifest imagery of the dream. Psychoanalysis seeks to translate the "disguised" manifest content into the true latent, and therefore repressed, wishes of the dreamer.

Latent learning Learning that occurs without apparent reinforcement but is not demonstrated until such time as reinforcement occurs.

Lateral fissure The visible fissure in the brain that separates the frontal lobe from the temporal lobe.

Lateral geniculate nucleus (LGN) Part of the visual pathway between the retina and the primary visual cortex.

Lateral hypothalamus A region of the hypothalamus which is some-times said to be a "hunger center" and to be in an antagonistic relation to a supposed "satiety center," the ventromedial region of the hypothalamus.

Lateral inhibition The suppression of the response in one neurone by the responses of other neurones dealing with nearby regions of the stimulus. A property of almost all sensory systems, it serves to filter out uniform stimulation and emphasize spatial changes.

Lateralization An asymmetry of function of the two cerebral hemispheres. In most right-handers, the left hemisphere is specialized for language functions, while the right hemisphere is better at various visual and spatial tasks.

Laughter therapy Making people laugh as means to relieve tension or depression.

Law of closure A Gestalt law of organization; elements missing from the outline of a figure are "filled in" by the visual system.

Law of common fate A Gestalt law of organization; elements that move together give rise to the perception of a particular figure.

Law of effect A theory that the tendency of a stimulus to evoke a response is strengthened if the response is followed by reward and is weakened if the response is not followed by reward. Applied to instrumental learning, this theory states that as trials proceed, incorrect bonds will weaken while the correct bond will be strengthened.

Law of proximity A Gestalt law of organization; elements located closest to each other are perceived as belonging to the same figure.

Law of similarity A law of grouping that states that the most similar elements are grouped together.

Laws of perceptual organization Max Wertheimer's formulation describing our predisposition to group stimuli based on their proximity, similarity, and good continuation.

Leader The person who exerts the most influence on group behaviour and beliefs.

Leadership The process by which certain group members motivate and guide the group.

Leading questions Questions asked during an investigation that are worded in such a manner that will suggest specific answers. This sort of questioning should be avoided and may become inadmissible in a court case.

Learned helplessness A condition created by exposure to inescapable aversive events. This retards or prevents learning in subsequent situations in which escape or avoidance is possible.

Learned helplessness theory of depression The theory that depression is analogous to learned helplessness effects produced in the laboratory.

Learned taste aversion A specialized form of learning in which an organism learns to avoid a taste after just one pairing of that taste with illness. For example, an animal will avoid a food that on an earlier occasion made it sick.

Learning Learning is the process of acquiring knowledge, skills, attitudes, or values, through study, experience, or teaching, that causes a change of behaviour that is persistent, measurable, and specified or allows an individual to formulate a new mental construct or revise a prior mental construct (conceptual knowledge such as attitudes or values). It is a process that depends on experience and leads to long-term changes in behaviour potential. Behaviour potential describes the possible behaviour of an individual (not actual behaviour) in a given situation in order to achieve a goal. But potential is not enough; if individual learning is not periodically reinforced, it becomes shallower and shallower, and eventually is lost in that individual.

Learning by insight This is learning by cognitive reorganization or reorganization of elements of the perceptual field. This occurs all on a sudden and this shows that even lower animals are capable of higher types of learning when given the circumstances which allow the operation of higher forms of learning.

Learning curve A curve in which some index of learning (e.g., the number of drops of saliva in Pavlov's classical conditioning experiment) is plotted against trials or sessions.

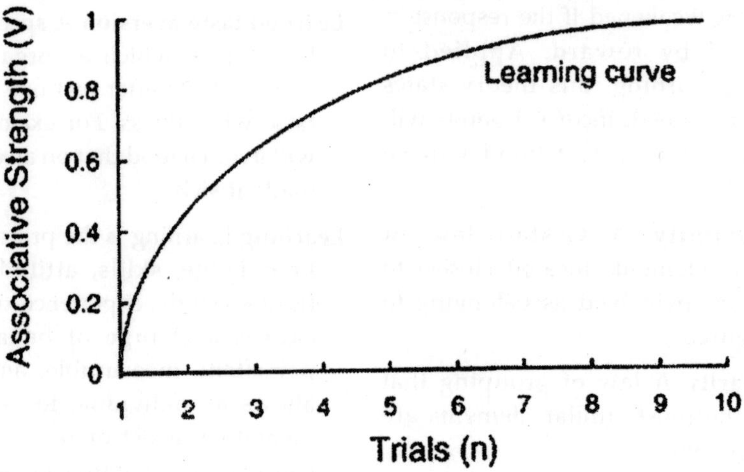

Fig. Learning curve

Learning difficulties The various ways in which children underachieve or perform badly at school, often caused by dyslexia o other difficulties in dealing with reading, writing or solving mathematical problems.

Learning disabilities (LD) Disorders that impede academic progress of people who are not mentally retarded or emotionally disturbed.

Learning goals A motivational orientation of students who place primary emphasis on knowledge acquisition and self-improvement.

Learning model As defined in the text, a subcategory of the pathology model that 1. views mental disorders as the result of some form of faulty learning, and

2. Believes that these should be treated by behaviour therapists according to the laws of classical and instrumental conditioning or by cognitive therapists who try to affect faulty modes of thinking.

Learning model of addiction The idea that addiction is a way of coping with stress.

Learning objectives Specific behaviours students are expected to exhibit at the end of a series of lessons.

Learning-performance distinction The difference between what has been learned and what is expressed in overt behaviour.

Learning probe Methods, such as questions, that help teachers find out if students understand a lesson.

Learning set The increased ability to solve various problems, especially in discrimination learning, as a result of previous experience with problems of a similar kind.

Learning styles Orientation for approaching learning tasks and processing information in certain ways.

Learning theory Based on the idea that changes in behaviour result more from experience and less from our personality or how we think or feel about a situation.

Learning together A cooperative learning model that involves students with four- or five-member heterogenous groups on assignments.

Least preferred co-worker (LPC) In fiedler's contingency theory the identification of a fellow group-member on a scale indicating how far the leader can separate the personal characteristics of this person from the work characteristics.

Least restrictive environment Provisions in the law (IDEA) that requires students with disabilities to be educated to the maximum extent appropriate with their nondisabled peers.

Legal authoritarianism A juror whom tends to assume the worst about an accused defendant, and is found more likely to convict. See "leniency bias".

Legibility of places In environmental cognition, the image people develop of places in a form which is easy to comprehend and to remember.

Legitimate power Power derived through one's position, such as a police officer or elected official.

Leniency bias A juror whom tends to make favourable assumptions regarding the accused. See "legal authoritarianism".

Lens The portion of the eye that bends light rays and thus can focus an image on the retina.

Leprophobia or lepraphobia Fear of leprosy.

Leptin A chemical produced by the adipose cells that seems to signal that plenty of fat is stored and that no more fat is needed. This signal may diminish eating.

Lesioning studies An invasive method of studying the brain in which a specific, small area of the brain is destroyed.

Lesions Injuries to or destruction of brain tissue.

Lesson planning Involves stating learning objectives; thinking through what the students will know or be able to do after the lesson; what information, activities, and experiences the teacher will provide; the time needed to reach the objective; what books, materials, and media support will be provided by the teacher; and the method(s) of instruction.

Leukophobia Fear of the colour white.

Level An aspect of a variable—e.g. "male" and "female" if the variable is gender.

Levels-of-processing theory A theory that suggests that the deeper the level at which information was processed, the more likely it is to be retained in memory.

Levophobia Fear of things to the left side of the body.

Lexical access The process of recognizing and understanding a word, which is presumably achieved by making contact with (accessing) the word in the mental lexicon.

Lexical category Also known as parts of speech, this is a class of words which appear to share some of the same features.

Lexical decision task A task in which the participant must decide as quickly as possible whether a stimulus (*book, trup, filt*) is or is not a word.

Libido a) In Freud's theory, an emotional and psychic energy derived from the biological drive of sexuality. b) In Jung's theory, an undifferentiated life and psychic energy. An insistent, instinctual force that is unresponsive to the demands of reality; the primary source of motivation. Freud's term for sexual energy, the driving force of human behaviour. In his later writings he used the term more diffusely to cover all kinds of psychic energy. Generally speaking the term is currently used to denote energy whose origins at least are sexual and which is dynamic and irrational in expression. Psychic energy-freud spoke of libido as just sexual but jung regards libido as all psychic energy. Sigmund Freud's terminology of sexual energy or sexual drive. The psychic drive or energy usually associated with the sexual instinct. (Sexual is used here in the broad sense to include pleasure and love-object seeking.) psychological energy (don't confuse with Freud's "libido," an inherently sexual drive energy) that is finalistic and founded not on substances but their relations and movements. Always in advance of consicousness, calling us into new activity. Libido in turn is a part of the life energy that drives all organisms to grow and develop. Its first expression is in the energy of growth that causes cell division, budding, etc. (so it is sexual at first). As you climb the phylogenetic ladder, libido used for sexuality loses its sexual character and flows into other forms.

Libido contains two opposite urges or instincts to live and to die, to go forwards and backwards (death drive) into instinctuality/womb/uncon. The libido contains both or no movement could happen. In youth the life instinct is stronger. It's not split; it just flows between the two poles. All energy flows from a difference in potentials. (Libido also splits into endogamous and exogamous tendencies. Incestuous libido connected to mother and family must be directed outward or it will remain fixed in the incestuous bond. Mysteries and their symbols reapply this libido productively. Work makes true individuality possible.

Some important ideas:.

— The principle of equivalence: energy disappearing from one psychological content will appear in another (similar to the first law of thermodynamics). When energy disappears, it must invest an unconscious content, and it will carry along features of the system losing it.

— Highly developed systems tend to seize energy from others.

— Energy flows down a gradient from highly charged inner realities to those with a lower charge (similar to the second law of thermodynamics, which describes entropy).

— Energy regresses when it flows out of some system, depotentiating the opposites and adapting to the unconscious, and progresses when it flows in to enhance outer adaptation (adaptation has two stages: attainment of attitude and completing adaptation by means of the attitude)(note: intro- and extraversion can be both). Infantile regression is more permanent and aims at incest and nourishment; mature regression avoids this, fasts instead, and its libido switches over to a symbol or symbolic equivalent of the "alma mater," the collective unconscious. But the psyche doesn't succumb to entropy because it's a relatively closed energy system.

— The whole psyche tries to maintain an energic balance but sometimes fails temporarily, as shown by important contents that, deprived of energy, slip into unconsciousness.

— Aspects of the outer world and the archetypes act as power sources that resupply the psyche.

— The polarity between the instincts and their archetypal images creates the gradiant that makes psychological energy possible. The polarity is like that between the terminals of a charged battery. Drives are specific energy manifestations.

Some libido symbols: energy, fire, sun, magic, gods and goddesses, electricity, sex, fertility, potency. The psychosexual energy originating in the id. Libido is the electric current of the mechanism of personality. It powers all psychological operations, invests desires, and undergoes ready displacement. It is the basic fuel of the self. Because it is of a relatively fixed quantity, like gasoline in a tank, it obeys laws of psychical "economy" in that a surplus in one system means a loss somewhere else. It can be either free or bound.

Lie scales Statistics used to provide information about the likelihood that a subject is lying in a test.

Life-change units (LCUs) In stress research, the measure of the stress levels of different types of change experienced during a given period.

Life cycle The various stages of life and development that a person experiences from birth to death, which some psychologists are now emphasising in studies of the life-span development perspective in adults.

Life review The process of reminiscing and examining the scope of one's life.

Life span development The study of the continuities, stabilities, and changes in psychological and physical processes that characterize human functioning, from conception through the final phases of life.

Light A kind of electromagnetic radiation emitted by the sun, stars, fire, and lightbulbs.

Light adaptation The process by which receptor cells become less sensitive to light.

Light intensity The amount of light emitted or reflected by an object.

Lightness constancy The tendency to perceive the lightness of an object as more or less the same despite the fact that the light reflected from the object changes with the illumination that falls upon it.

Lightness contrast The perceptual phenomenon by which the lightness of one region of the image is affected by nearby regions, such as how a given region will look darker when surrounded by a bright region than when surrounded by a light region.

Ligyrophobia Fear of loud noises.

Likeability In environmental psychology, the qualities accounting for preferences of place and location.

Liking A feeling of personal regard, intimacy, and esteem toward another person.

Lilapsophobia Fear of tornadoes and hurricanes.

Limbic cortex The cerebral cortex located around the edge of the cerebral hemispheres where they join with the brain stem; part of the limbic system.

Limbic system A brain system that plays a role in emotional expression, particularly in the emotional component of behaviour, memory, and motivation.

Limited English proficiency (LEP) Descriptive term for students who have limited mastery of English.

Limnophobia Fear of lakes.

Line of best fit A line drawn through the points in a scatter diagram. It yields the best prediction of one variable when given the value of the other variable.

Linear perspective One of the pictorial cues to depth which relies on the fact that objects appear smaller when they are further from the viewer. As a result, parallel lines receding into the distance seem to converge, and the eye uses this pattern as a cue to distance.

Linguistic determinism The theoretically 'stronger' version of the sapir-whorf hypothesis, that language must determine thought, i.e. That what one says causally affects what one thinks.

Linguistic relativity A theoretically 'weak' version of the sapir-whorf hypothesis that language must influence thought, but which does not take account of the fact that if a language has no word for something, then it cannot be thought about.

Link method The process of associating items with one another in order to remember them.

Linkage In the study of genetics, the link between the genetic basis of characters which results from both genes lying on the same chromosome.

Linonophobia Fear of string.

Literacy skills The acquisition of grammar leading to the ability to read and write.

Lithium A drug prescribed for treating bipolar disorders.

Lithium carbonate A drug used in the treatment of mania and bipolar disorders.

Liticaphobia Fear of lawsuits.

Living will A document, legal in some states, directing that all measures to support life be ended if the signer should be dying of an incurable condition.

Lobotomy A surgical procedure that severs nerve tracts in the frontal lobe, formerly used to treat certain psychological disorders but now rarely performed.

Local optimality In the study of evolution, the optimization by natural selection on a local scale, of animals living within a particular habitat.

Local representations A model of cognitive organization, especially semantic memory, in which each concept is represented by a single node or, more plausibly, a group of nodes. *See also* distributed representations, node.

Localization of function The process of determining what each region of the brain contributes to which aspects of thinking and behaviour.

Loci method Strategy for remembering lists by picturing items in familiar locations.

Lock-and-key model The theory that neurotransmitter molecules will only affect the postsynaptic membrane if their shape fits into that of certain synaptic receptor molecules.

Locke, John The first great British empiricist, denied the existence of innate ideas, categories, and moral principles. The mind at birth is a tabula rasa. Its whole content is derived from sense-experience, and constructed by reflection upon sensible data. Reflection is effected through memory and its attendant activities of contemplation, distinction, comparison in point of likeness and difference, and imaginative recompositon. Even the most abstract notions and ideas, like infinity, power, cause and effect, substance and identity, which seemingly are not given by experience, are no exceptions to the rule. Thus "infinity" confesses our inability to limit in fact or imagination the spatial and temporal extension of sense-experience; "substance," to perceive or understand why qualities congregate in separate clumps; "power" and 'cause and effect," to perceive or understand why and how these clumps follow, and seemingly produce one another as they do, or for that matter, how our volitions "produce" the movements that put them into effect. Incidentally, Locke defines freedom as liberty, not of choice, which is always sufficiently motivated, but of action in accordance with choice. "Identity" of things, Locke derives from spatial and temporal continuity of the content of clumps of sensations; of structure, from continuity of arrangement in changing content; of person, from continuity of consciousness through memory, which, incidentally, permits of alternating personalities in the same body or of the transference of the same personality from one body to another.

In these circumstances real knowledge is very limited. "Universals" register superficial resemblances, not the real essences of things. Experience directly "intuits" identity and diversity, relations, coexistences and necessary connections in its content, and, aided by memory, "knows" the agreements and disagreements of ideas in these respects. We also feel directly (sensitive knowledge) that our experience comes from without. Moreover, though taste, smell, colour, sound, etc. are internal to ourselves (secondary qualities) extension, shape, rest, motion, unity and plurality (primary qualities) seem to inhere in the external world independently of our perception of it. Finally, we have "demonstrative knowledge" of the existence of God. But of anything other than God, we have no knowledge except such as is derived from and limited by the senses.

Lockiophobia Fear of childbirth.

Locus of control People tend to ascribe their chances of future successes or failures either to internal or external causes. Persons with an internal locus of control see themselves as responsible for the outcomes of their own actions. These individuals often believe that they control their destiny, and are often observed to excel in educational or vocational realms. Someone with an external locus of control, on the other hand, sees environmental causes and situational factors as being more important than internal ones. These individuals would be more likely to see luck rather than effort as determining whether they succeed or fail in the future, and are more likely to view themselves as the victim in any given situation.

Locution In speech act theory, the literal speech act, i.e. The utterance itself.

Logic The general science of inference. Deductive logic, in which a conclusion follows from a set of premises, is distinguished from inductive logic, which studies the way in which premises may support a conclusion without entailing it. In deductive logic the conclusion cannot be false if the premises are true. The aim of a logic is to make explicit the rules by which inferences may be drawn, rather than to study the actual reasoning processes that people use, which may or may not conform to those rules. In the case of deductive logic, if we ask why we need to obey the rules, the most general form of answer is that if we do not we contradict ourselves (or, strictly speaking, we stand ready to contradict ourselves. Someone failing to draw a conclusion that follows from a set of premises need not be contradicting him or herself, but only failing to notice something. However, he or she is not defended against adding the contradictory conclusion to his or her set of beliefs.) There is no equally simple answer in the case of inductive logic, which is in general a less robust subject, but the aim will be to find reasoning such that anyone failing to conform to it will have improbable beliefs. Aristotle is generally recognized as the first great logician, and Aristotelian logic or traditional logic dominated the subject until the 19th century. It has become increasingly recognized in the 20th century that fine work was done within that tradition, but syllogistic reasoning is now generally regarded as a limited special case of the forms of reasoning that can be represented within the propositional and predicate calculus. These form the heart of modern logic. Their central notions, of quantifiers, variables, and functions were the creation of the German mathematician Frege, who is recognized as the father of modern logic, although his treatment of a logical system as an abstract mathematical structure, or algebra, had been heralded by Boole. Modern logic is thus called mathematical logic for two reasons: first, the logic itself is an object of mathematical study, but secondly, the forms introduced by Frege provided a language capable of representing all mathematical reasoning. This was something traditional logic had been quite incapable of tackling. The propositional and predicate calculus study ways of combining propositions with the connectives expressing truth-functions, and of combining information about the quantity of times predicates are satisfied. These highly general operations can occur in any discourse, from mathematics to discussion of the football results. More specific logics study particular topics such as time, possibility, and obligation. Thus there exist deontic logics, modal logics, logics of tense, and so on. For other notions associated with the study of logic interpretation, logical calculus, logical constants, logical form, model theory, proof theory, quantifier, truth-function, variable.

Logistic system The formal construction of a logistic system requires: 1. a list of primitive symbols (these are usually taken as marks but may also be sounds or other things—they must be capable of instances which are, recognizably, the same or different symbols, and capable of utterance in which instances of them are put forth or arranged in an order one after another);

2. A determination of a class of formulas, each formula being a finite sequence of primitive symbols, or, more exactly, each formula being capable of instances which are finite sequences of instances of primitive symbols (generalizations allowing two-dimensional arrays of primitive symbols and the like are non-essential);

3. A determination of the circumstances under which a finite sequence of formulas is a proof of the last formula in the sequence, this last formula being then called a theorem (again we should more exactly speak of proofs as having instances which are finite sequences of instances of formulas);

4. A determination of the circumstances under which a finite sequence of formulas is a proof of the last formula of the sequence as a consequence of a certain set of formulas (when there is a proof of a formula B as a consequence of the set of formulas A 1, A 2, . . . , A n, we say that the inference from the premisses A 1, A 2, . . . , An to the conclusion B is a valid inference of the logistic system). It is not excluded that the class of proofs in the sense of (3) should be empty. But every proof of a formula B as a consequence of an empty-set of formulas, in the sense of (4), must also be a proof of B in the sense of (3), and conversely. Moreover, if to the proof of a formula B as a consequence of A 1, A 2, . . . An are prefixed in any order proofs of A 1, A 2,..., An, the entire resulting sequence of formulas must be a proof of B; more generally, if to the proof of a formula B as a consequence of A 1, A 2, . . . , An are prefixed in any order proofs of a subset of A 1, A 2, . . . , An as consequences of the remainder of A 1, A 2, . . . , An, the entire resulting sequence must be a proof of B as a consequence of this remainder.

The determination of the circumstances under which a sequence of formulas is a proof, or a proof as a consequence of a set of formulas, is usually made by means of: (5) a list of primitive formulas; and (6) a list of primitive rules of inference each of which prescribes that under certain circumstances a formula B shall be an immediate consequence of a set of formulas A 1, A 2, . . . , An. The list of primitive formulas may be empty—this is not excluded. Or the primitive formulas may be included under the head of primitive rules of inference by allowing the case n=0 in (6). A proof is then defined as a finite sequence of formulas each of which is either a primitive formula or an immediate consequence of preceding formulas by one of the primitive rules of inference. A proof as a consequence of a set of formulas A 1, A 2, . . . , An is in some cases defined as a finite sequence of formulas each of which is either a primitive formula, or one of A 1, A 2, . . . , An, or an immediate consequence of preceding formulas by one of the primitive rules of inference; in other cases it may be desirable to impose certain restrictions upon the application of the primitive rules of inference (e.g., in the case of the functional calculus of first order—logic, formal, 3—that no free variable of A1, A 2, . . . , An shall be generalized upon).

A logistic system need not be given any meaning or interpretation, but may be put forward merely as a formal discipline of interest for its own sake; and in this case the words proof, theorem, valid inference, etc., are to be dissociated from their everyday meanings and taken purely as technical terms. Even when an interpretation of the system is intended, it is a requirement of rigor that no use shall be made of the interpretation (as such) in the determination whether a sequence of symbols is a formula, whether a sequence of formulas is a proof, etc.

The kind of an interpretation, or assignment of meaning, which is normally intended for a logistic system is indicated by the technical terminology employed. This is namely such an interpretation that the formulas, some or all of them, mean or express propositions; the theorems express true propositions; and the proofs and valid inferences represent proofs and valid inferences in the ordinary sense.A logistic system may thus be regarded as a device for obtaining—or, rather stating—an objective, external criterion for the validity of proofs and inferences (which are expressible in a given notation).

A logistic system which has an interpretation of the kind in question may be expected, in general, to have more than one such interpretation.

It is usually to be required that a logistic system shall provide an effective criterion for recognizing formulas, proofs, and proofs as a consequence of a set of formulas; i.e., it shall be a matter of direct observation, and of following a fixed set of directions for concrete operations with symbols, to determine whether a given finite sequence of primitive symbols is a formula, or whether a given finite sequence of formulas is a proof, or is a proof as a conseuqence of a given set of formulas. If this requirement is not satisfied, it may be necessary —e.g.—given a particular finite sequence of formulas, to seek by some argument adapted to the special case to prove or disprove that it satisfies the conditions to be a proof (in the technical sense); i.e., the criterion for formal recognition of proofs then presupposes, in actual application, that we already know what a valid deduction is (in a sense which is stronger than that merely of the ability to follow concrete directions in a particular case).

The requirement of effectiveness does not compel the lists of primitive symbols, primitive formulas, and primitive rules of inference to be finite. It is sufficient if there are effective criteria for recognizing formulas, for recognizing primitive formulas, for recognizing applications of primitive rules of inference, and (if separately needed) for recognizing such restricted applications of the primitive rules of inference as are admitted in proofs as a consequence of a given set of formulas.

With the aid of Gdel's device of representing sequences of primitive symbols and sequences of formulas by means of numbers, it is possible to give a more exact definition of the notion of effectiveness by making it correspond to that of recursiveness of numerical functions. E.g., a criterion for recognizing primitive formulas is effective if it determines a general recursive monadic function of natural numbers whose value is 0 when the argument is the number of a primitive formula, 1 for any other natural number as argument. The adequacy of this technical definition to represent the intuitive notion of effectiveness as described above is not immediately clear, but is placed beyond any real doubt by developments for details of which the reader is referred to Hilbert-Bernays and Turing.

The requirement of effectiveness plays an important rle in connection with logistic systems, but the necessity of the requirement depends on the purpose in hand and it may for some purposes be abandoned. Various writers have proposed non-effective, or non-constructive, logistic systems; in some of these the requirement of finiteness of length of formulas is also abandoned and certain infinite sequences of primitive symbols are admitted as formulas.

Logizomechanophobia Fear of computers.

Logophobia Fear of words.

Loneliness An emotional state resulting from the desire for close relationships but being unable to attain them. This is a common attribute of sex offenders.

Long range motion system One of two systems thought to underpin human motion perception which operates indirectly, by inferring motion from the change in image position of a given feature over time.

Long term memory A type of memory in which information is stored indefinitely.

Longitudinal design A research design in which the same participants are observed repeatedly, sometimes over many years.

Longitudinal fissure The front-to-back cleavage that divides the left and right hemispheres of the brain.

Longitudinal study In personality assessment, a study of adult development drawn from research data compiled from the same group of adults at different points in their lives, which are based on material form q- and r-data.

Long-term memory A memory system that stores an unlimited amount of information permanently.

Long-term potentiation (LTP) A form of cellular plasticity in which a postsynaptic neuron becomes more sensitive (potentiated) to the signal received from the presynaptic neuron. This potentiation is usually produced by a rapid and sustained burst of firing by the presynaptic neuron. The potentiation can then spread to other presynaptic neurons provided that they have fired in the past at the same time as the presynaptic cell that produced the potentiation in the first place.

Long-term store (LTS) The permanent store of knowledge which is of infinite capacity and which is often referred to as secondary memory.

Looming A rapid magnification of a form in the visual field that generally signals impending impact.

Loss aversion A widespread pattern, evident in many aspects of decision making, in which people seem particularly sensitive to losses and eager to avoid them. In many cases, this manifests itself as an increased willingness to take risks in hopes of reducing the loss.

Loudness A perceptual dimension of sound influenced by the amplitude of a sound wave; sound waves with large amplitudes are generally experienced as loud and those with small amplitudes as soft.

Loving A combination of liking and a deep sense of attachment to, intimacy with, and caring for another person.

Low birthweight baby A baby born weighing less than 5-1/2 pounds.

Low-ball technique A tactic for getting people to agree to something. People who agree to an initial request will often still comply when the requester ups the ante. People who receive only the costly request are less likely to comply with it.

Low-balling A two-step compliance technique in which the influencer secures agreement with a request but then increases the size of that request by revealing hidden costs.

Lower confidence limit The lower limit of a confidence interval. If prediction states that the true score falls between 80 and 90, then the lower confidence level is 80.

Lucid dreaming The theory that conscious awareness of dreaming is a learnable skill that enables dreamers to control the direction and content of their dreams.

Luiphobia Fear of lues, syphillis.

Luminance ratio The ratio between the light reflected off a figure and that reflected off the background against which the figure is seen. Whatever the illumination, the ratio remains the same.

Lunar month A four-week period of 28 days.

Lutraphobia Fear of otters.

Lygophobia Fear of darkness.

Lyssophobia Fear of rabies or of becoming mad.

M

Mach bands The accentuated edges between two adjacent regions that differ in brightness. This sharpening is maximal at the borders where the distance between the two regions is smallest and the contrast most striking.

Machiavellian intelligence In the theory of evolution, the development of an intelligent response to social complexity, where animals living in competitive social groups learn to manipulate, deceive and to cooperate according to the dictates of circumstance.

Macrophobia Fear of long waits.

Mageirocophobia Fear of cooking.

Magic number According to George Miller, the number (seven plus or minus two) that represents the holding capacity of the working memory system.

Magnetic Resonance Imaging (MRI) A technique for exploring the human brain, using magnetic fields and radio waves to generate and record energy pulses within the brain.

Magnitude Characteristic of a scale of measurement where the individual units possess the qualities of greater than, equal to, or less than.

Magno cells Ganglion cells found largely in the periphery of the retina that, because of their sensitivity to brightness changes, are particularly suited to the perception of motion and depth.

Maieusiophobia Fear of childbirth.

Main effect A statistical term indicating the overall effect that an independent variable has on the dependent variable, ignoring all other independent variables.

Mainstreaming The placement, for all or part of the school day, of disabled children in regular classes.

Maintenance Continuation of behaviour.

Maintenance rehearsal Repetition to keep material in working memory for a while. In contrast to elaborative rehearsal, this confers little long-term benefit.

Major depression A mood disorder in which patients are disabled by guilt or sadness (especially in Western cultures), experience a loss of energy, pleasure and motivation, and disturbances of sleep, diet, and other bodily functions.

Major depressive disorder A mood disorder characterized by intense feelings of depression over an extended time, without the manic high phase of bipolar depression.

Major depressive episode A period of at least two weeks marked by sadness or irritability and loss of interest in activities. Other symptoms may include changed sleeping or eating patterns, low energy, feelings of worthlessness or guilt, difficulty concentrating, and recurrent thoughts about suicide.

Malaxophobia Fear of love play. (Sarmassophobia).

Male climacteric Male menopause.

Malimprinting When the right object for imprinting is not available the organism attaches itself to the object nearest to the search image.

Managed care An arrangement for health care in which an organization, such as a health maintenance organization, acts as an intermediary between a person seeking care and a treatment provider.

Mania Psychotic state characterized by over-activity, overconfidence, excitement, delusions, etc.

Maniaphobia Fear of insanity.

Manic depressive Psychotic cyclothymic temperament characterized by frequent oscillations of mood (over-activity vs withdrawal).

Manic episode A component of bipolar disorder characterized by periods of extreme elation, unbounded euphoria without sufficient reason, and grandiose thoughts or feelings about personal abilities.

Manic state Uncontrolled excitement, restlessness, overactivity, explosive superficial happiness, overconfidence.

Manifest content According to Freud, the story-like superficial content of a dream, often representing only the daily activities and little underlying unconscious material.

Mapping Diagramming main ideas and connections between them.

Marriage To be joined as husband and wife; united in wedlock.

Masculinity Possession of instrumental personality traits.

Masking Attempting to hide the expression of an emotion.

Maslow, Abraham (1908-1970) An American personality theorist and leading exponent of humanistic psychology, closely associated with the terms peak experience and self-actualisation. "i was awfully curious to find out why i didn't go insane," remarked abraham maslow, one of the founders of humanistic psychology.

He was born and raised in Brooklyn, the eldest of seven children. He was smart but shy, and remembered his childhood as lonely and rather unhappy. Maslow attended City College in New York. His father hoped he would pursue law, but he went to graduate school at the University of Wisconsin to study psychology. While there, he married his cousin Bertha, and found as his chief mentor Professor Harry Harlow. At Wisconsin he pursued an original line of research, investigating primate dominance behavior and sexuality. He went on to further research at Columbia University, continuing similar studies. He found another mentor in Alfred Adler, one of Freud's early followers.

From 1937 to 1951, Maslow was on the faculty of Brooklyn College. In New York he found two more mentors, anthropologist Ruth Benedict and Gestalt psychologist Max Wertheimer, whom he admired both professionally and personally. These two were so accomplished in both realms, and such "wonderful human beings" as well, that Maslow began taking notes about them and their behavior. This would be the basis of his lifelong research and thinking about mental health and human potential. He wrote extensively on the subject, borrowing ideas from other psychologists but adding significantly to them, especially the concepts of a heirarchy of needs, metaneeds, self-actualizing persons, and peak experiences. Maslow became the leader of the humanistic school of psychology that emerged in the 1950s and 1960s, which he referred to as the "third force" – beyond Freudian theory and behaviorism.

Maslow saw human beings' needs arranged like a ladder. The most basic needs, at the bottom, were physical – air, water, food, sex. Then came safety needs – security, stability – followed by psychological, or social needs – for belonging, love, acceptance. At the top of it all were the self-actualizing needs – the need to fulfill oneself, to become all that one is capable of becoming. Maslow felt that unfulfilled needs lower on the ladder would inhibit the person from climbing to the next step. Someone dying of thirst quickly forgets their thirst when they have no oxygen, as he pointed out. People who dealt in managing the higher needs were what he called self-actualizing people. Benedict and Wertheimer were Maslow's models of self-actualization, from which he generalized that, among other characteristics, self-actualizing people tend to focus on problems outside of themselves, have a clear sense of what is true and what is phony, are spontaneous and creative, and are not bound too strictly by social conventions.

Peak experiences are profound moments of love, understanding, happiness, or rapture, when a person feels more whole, alive, self-sufficient and yet a part of the world, more aware of truth, justice, harmony, goodness, and so on. Self-actualizing people have many such peak experiences.

Maslow's thinking was surprisingly original – most psychology before him had been concerned with the abnormal and the ill. He wanted to know what constituted positive mental health. Humanistic psychology gave rise to several different therapies, all guided by the idea that people possess the inner resources for growth and healing and that the point of therapy is to help remove obstacles to individuals' achieving this. The most famous of these was client-centered therapy developed by Carl Rogers.

Maslow was a professor at Brandeis University from 1951 to 1969, and then became a resident fellow of the Laughlin Institute in California. He died of a heart attack in 1970.

Masochism Caused by an alloy of erotism and a sadistic superego. It fills the ego with a desire to atone by punishing itself. An original or primary masochism develops into a 1. feminine,

2. Erotogenic, or

3. Moral masochism via a turning of unresolved sadism against oneself (secondary masochism). Too much internalized death drive and you get masochism; when directed outward, sadism.

Massed practice Technique in which facts or skills to be learned are repeated many times over a concentrated period of time.

Master status A socially defined position occupied by a person in society that is very important in shaping his or her self-concept and life choices.

Mastery criterion A standard students must meet to be considered proficient in a skill.

Mastery goals The goals students must reach to be considered proficient in a skill.

Mastery grading Absolute grading based on criteria for mastery.

Mastery learning System of instruction that emphasizes the achievement of instructional objectives by all students by allowing learning time to vary.

Mastigophobia Fear of punishment.

Matching A systematic selection of subjects in groups in an experiment or (more often) a correlational study to ensure that the mean values of important subject variables of the groups are similar.

Matching bias model In analysis of poor performance in the selection task, a proposal that claims that subjects choose cards in the task in two ways, either by correct logical reasoning, or by looking at items in an abstract fashion.

Matching hypothesis The proposition that people are attracted to and form relationships with those who are similar to them in particular characteristics, such as physical attractiveness.

Matching items Given two lists, each item in one list will match with one item in the other list.

Matching phenomenon The tendency for men and women to choose as partners those who are a "good match" in attractiveness and other traits.

Matching to sample A procedure in which an organism has to choose which of two alternative stimuli is the same as a third, sample stimulus.

Mate guarding A male strategy of insuring paternity by remaining near the female and preventing her from mating with other males.

Materialism A philosophical belief that reality can be known only through an understanding of the physical world of which the mind is a part.

Maternal aggression Aggressive behaviour put forward by females (and most likely males as well) when an intruder is in the presence of ones children.

Maternal deprivation hypothesis A term defined by the english psychiatrist john bowlby to describe the critical period of around 6 months to 3 years during which the child needs continuous love and care from one person, the mother or a permanent mother-substitute.

Maturation A genetically programmed growth process that is relatively unaffected by environmental conditions (e.g., flying in sparrows and walking in humans).

Maturity The stage in human development when the adult is seen as having achieved certain socially-acceptable levels of sexual, intellectual and emotional fullness. As defined by erik erikson, the adult at this stage develops the need to concern him/herself with the development of the next generation, either through parenthood or through the passing on of skills and knowledge to others.

Maximization A questioning technique in which the interrogator exaggerates the strength of evidence gathered in order to elicit a confession. See 'minimization'.

Maximum-likelihood principle The assertion that we interpret the proximal stimulus pattern as that external stimulus object that most probably produced it.

M-cell A type of retinal ganglion cell which typically predominates in the central region of the retina, with a large receptive field and a transient response.

Mean A measure of central tendency determined by adding all scores together and dividing by the number of scores. Often referred to as the statistical average.

Meaningful learning Mental processing of new information leading to its linkage with previously learned knowledge.

Means end analysis A problem-solving technique that encourages identifying the goal (ends) of a problem, the current situation, and what needs to be done (means) to reduce the difference between the two conditions.

Measure of central tendency A single number intended to summarize an entire distribution of experimental results. Three commonly used measures of central tendency are: 1. the mode, or the score that occurs most frequently;.

2. The median, or the point that divides the distribution into two equal halves; and.

3. The mean, or the arithmetic average.

Measurement, scales of Categories of data based on their numerical characteristics.

Measures of variability A statistic, such as a range or standard deviation, that indicates how tightly the scores in a set of observations cluster together.

Mechanophobia Fear of machines.

Medial forebrain bundle (MFB) A tract of fibres that runs through the base of the forebrain and parts of the hypothalamus. Electric stimulation of this bundle is usually experienced as rewarding.

Median The number in a range of scores that falls exactly in the middle such that 50% of the cases are above and below it. Requires interval or ratio data.

Mediated learning A teaching method in which the teacher guides instruction so that students will master and internalize the skills that permit higher cognitive functioning.

Mediation An attempt by a neutral third party to resolve a conflict by facilitating communication and offering suggestions.

Medical model As defined in the text, a subcategory of the pathology model that holds 1. that the underlying pathology is organic and

2. That the treatment should be conducted by physicians.

Meditation A form of consciousness alteration designed to enhance self-knowledge and well-being through reduced self-awareness.

Medomalacuphobia Fear of losing an erection.

Medorthophobia Fear of an erect penis.

Medulla Part of the hindbrain and the rearmost portion of the brain, just adjacent to the spinal cord. It is involved in the control of respiration, circulation, balance, and protective reflexes such as coughing and sneezing.

Megalophobia Fear of large things.

Meiosis The form of cell division by which new sperm and ova are formed. The chromosomes within the cell are randomly rearranged so that new sperm and ova contain 23 individual chromosomes, or half of that found in other cells of the body.

Melancholia We now call this depression. For Freud it shares with mourning a gradual withdrawing of libido from an object known to be lost or dead—but it is different in that the unconscious hatred felt toward the object with which the ego narcissistically identifies is turned against the ego. One then achieves revenge against the lost object by getting depressed.

Melanophobia Fear of the colour black.

Melatonin A neurohormone secreted by the pineal gland that is involved in regulating the sleep-waking cycle.

Melissophobia Fear of bees.

Melophobia Fear or hatred of music.

Memory Memory is a function of the brain: the ability to retain information. Memory is much studied by cognitive psychology and neuroscience. There are multiple types of classifications for memory based on duration, nature and retrieval of perceived items. T.

Memory cueing hypothesis An explanation for the effects of content in the selection task, which suggests that familiar concrete content can remind a person of a similar problem that they have encountered in the past.

Memory effect Error in research that results from subjects recalling previous testing and applying that knowledge to current testing.

Memory reconstruction The automatic filling-in of gaps in memory, typically using general knowledge and expectations (e.g., remembering books on a professor's desk when in fact there were none).

Memory search A cognitive process preceding memory retrieval that usually occurs very quickly and outside our awareness.

Memory span The number of items that can be recalled after a single presentation.

Memory trace The physical basis of memory; a change in the nervous system produced by an experience. The exact nature of this change is still uncertain.

Menarche The first menstrual period of a girl in puberty.

Meninges The three layers of connective tissue that cover the brain and spinal cord. Metabolic costs in the study of evolution, the expenditure of metabolic energy by brain tissue in order to support intelligence.

Meningitophobia Fear of brain disease.

Menopause The permanent cessation of menstruation, normally between the ages of 40 and 50, or the period during which this occurs; female climacteric, or change of life.

Menophobia Fear of menstruation.

Mens rea The legal terminology referring to a perpetrators criminal mind. In order to be found guilty of a crime an individual must be proven to have acted within a criminal mind.

Menstrual flow The discharge consisting of the sloughed-off uterine lining that was built up in preparation for a fertilized ovum, that signals the onset of menstruation.

Menstrual synchrony The phenomenon wherein women who live together tend to develop menstrual cycles that roughly coincide with each other, even though their periods were very different at the outset.

Menstruate To have a menstrual period; undergo menstruation.

Mental age (MA) A score devised by Alfred Binet to represent a child's test performance. A child's MA is assessed by determining what mental tasks he can do sucessfully. If he can perform tasks generally performed well by six-year-olds but not tasks generally performed well by seven-year-olds, his MA is six, and so on.

Mental health The psychological state of someone who is functioning at a satisfactory level of emotional and behavioural adjustment.

Mental hospitals Medical institutions that specialize in providing treatment for psychological disorders.

Mental images Analogical representations that reserve some of the characteristic attributes of our senses.

Mental model A mental construction based on physical reality that is used to solve problems of logical deduction.

Mental models theory A framework for presenting research in thinking and reasoning based on manipulation of the models of parts of the world that people have in their minds.

Mental representations Internal symbols that stand for something but are not equivalent to it, such as internalized actions, images, or words.

Mental retardation Condition in which individuals have IQ scores 70 to 75 or below and also demonstrate limitations in the ability to bring adaptive skills to bear on life tasks.

Mental rotation A task in which participants are presented with a rotated figure and must discern whether the figure is normal or, say, mirror-reversed. Participants apparently must visualize the figure rotated to an upright position before responding.

Mental set A tendency to use only solutions that have worked in the past.

Mental space (M-Space) A hypothetical construct in Case's model of cognitive development similar to working memory, whose primary function is to process information from the external world.

Mere exposure effect The formation of a positive attitude toward a person, place, or thing based solely on repeated exposure to that person, place, or thing.

Mere presence theory The proposition that the mere presence of others is sufficient to produce social facilitation effects.

Merinthophobia Fear of being bound or tied up.

Mesoderm The middle layer of cells of an embryo, from which the skeletal, reproductive, muscular, vascular, and connective tissues develop.

Mesokurtic A curve or distribution that has a balanced amount of variance so that is resembles a normal curve.

Meta analysis The statistical procedure used to combine numerous and independent research results into one study. Each research study becomes one subject in the meta-analysis.

Meta motivation Maslow's term for selfless, detached motivation arising out of fullness and not deficiency.

Metacognition A general term for knowledge about knowledge, as in knowing that we do or don't remember something.

Metacognitive skills Methods for learning. studying. or solving problems.

Metalinguistic awareness The ability to recognize how people use language in its various linguistic forms and how language functions within particular cultures.

Metallophobia Fear of metal.

Metamemory The self-awareness of memory abilities in the young child which leads it to realize more about the need to apply strategies effectively.

Metaphysical mind In philosophical discussion of neuroscience, a definition of the human mind as being non-physical, i.e. As a separate entity that is not a property of brain function.

Metapsychological Taken together, those aspects of Freud's theorizing that are economical (the hydraulics of unpleasure-avoidance through pleasure), dynamic (libido movements among id, ego, and superego), and topographic (psyche as structured into conscious, preconscious, and unconscious layers). Metapsychology also takes clinical observations beyond the consulting room and applies them to everyone, with varying results. Although psychoanalysis began as a treatment method, Freud's real interest was caught by those theorizings that applied it to human psychology in general. Referring to two recent books, Freud wrote this to his friend Oskar Pfister:

"I do not know whether you have guessed the hidden link between my 'Lay Analysis' and 'Illusion.' In the former I want to protect analysis from physicians, and in the latter from priests. I want to entrust it to a profession that doesn't yet exist, a profession of secular ministers of souls, who don't have to be physicians and must not be priests.".

Metathesiophobia Fear of changes.

Meteorophobia Fear of meteors.

Method of loci A mnemonic technique that requires the learner to visualize each of the items she wants to remember in a different spatial location (locus). Recall requires that each location be mentally inspected for the item placed there.

Methodology The systematic analysis and organization of the rational and experimental principles and processes which must guide a scientific inquiry, or which constitute the structure of the special sciences more particularly. Methodology, which is also called scientific method, and more seldom methodeutic, refers not only to the whole of a constituted science, but also to individual problems or groups of problems within a science. As such it is usually considered as a branch of logic; in fact, it is the application of the principles and processes of logic to the special objects of the various sciences; while science in general is accounted for by the combination of deduction and induction as such. Thus, methodology is a generic term exemplified in the specific method of each science. Hence its full significance can be understood only by analyzing the structure of the special sciences. In determining that structure, one must consider (a) the proper object of the special science, (b) the manner in which it develops, (c) the type of statements or generalizations it involves, (d) its philosophical foundations or assumptions, and (e) its relation with the other sciences, and eventually its applications. The last two points mentioned are particularly important: methods of education, for example, will vary considerably according to their inspiration and aim. Because of the differences between the objects of the various sciences, they reveal the following principal methodological patterns, which are not necessarily exclusive of one another, and which are used sometimes in partial combination. It may be added that their choice and combination depend also in a large degree on psychological motives. In the last resort, methodology results from the adjustment of our mental powers to the love and pursuit of truth.

I. There are various rational methods used by the speculative sciences, including theology which adds certain qualifications to their use. More especially, philosophy has inspired the following procedures: 1. The Socratic method of analysis by questioning and dividing until the essences are reached;

2. The synthetic method developed by Plato, Aristotle and the Medieval thinkers, which involves a demonstrative exposition of the causal relation between thought and being;

3. The ascetic method of intellectual and moral purification leading to an illumination of the mind, as proposed by Plotinus, Augustine and the mystics;

4. The psychological method of inquiry into the origin of ideas, which was used by Descartes and his followers, and also by the British empiricists;

5. The critical or transcendental method, as used by Kant, and involving an analysis of the conditions and limits of knowledge;

6. The dialectical method proceeding by thesis, antithesis and synthesis, which is promoted by Hegelianism and Dialectical Materialism;

7. The intuitive method, as used by Bergson, which involves the immediate perception of reality, by a blending of consciousness with the process of change;

8. The reflexive method of metaphysical introspection aiming at the development of the immanent realities and values leading man to God;

9. The eclectic method (historical-Critical) of purposive and effective selection as proposed by Cicero, Suarez and Cousin; and
10. The positivistic method of Comte, Spencer and the logical empiricists, which attempts to apply to philosophy the strict procedures of the positive sciences.

II. The axiomatic or hypothetico-deductive method as used by the theoretical and especially the mathematical sciences. It involves such problems as the selection, independence and simplification of primitive terms and axioms, the formalization of definitions and proofs, the consistency and completeness of the constructed theory, and the final interpretation.

III. The nomological or inductive method as used by the experimental sciences, aims at the discovery of regularities between phenomena and their relevant laws. It involves the critical and careful application of the various steps of induction: observation and analytical classification; selection of similarities; hypothesis of cause or law; verification by the experimental canons; deduction, demonstration and explanation; systematic organization of results; statement of laws and construction of the relevant theory.

IV. The descriptive method as used by the natural and social sciences, involves observational, classificatory and statistical procedures and their interpretation.

V. The historical method as used by the sciences dealing with the past, involves the collation, selection, classification and interpretation of archeological facts and exhibits, records, documents, archives, reports and testimonies.

VI. The psychological method, as used by all the sciences dealing with human behaviour and development. It involves not only introspective analysis, but also experimental procedures, such as those referring to the relations between stimuli and sensations, to the accuracy of perceptions (specific measurements of intensity) to gradation (least noticeable differences), to error methods (average error in right and wrong cases), and to physiological and educational processes.

Methyphobia Fear of alcohol.

Metrophobia Fear or hatred of poetry.

Microbiophobia Fear of microbes. (Bacillophobia).

Microcircuitry Networks of interneurons within which most of the brain's information processing occurs.

Microexpressions A brief, incomplete, facial expression that occur on our faces very quickly after exposure to a stimulus. It occurs before we can actively conceal them. A trained observer may look for these to see what questions elicit certain responses during an interrogation or criminal trial.

Microgenesis In environmental psychology, the development of knowledge about a new environmental area based on the skills acquired through ontogenesis.

Microphobia Fear of small things.

Mid life crisis A personal identity conflict that comes during middle adulthood, usually around one's forties. The result of feelings of worthlessness and unresolved problems with intimacy and identity; frequently manifested in self-indulgent acts reflective of adolescent behaviour.

Midbrain The part of the brain between the hindbrain and forebrain that is involved in locating events in space and that contains a dopamine-releasing system of neurons.

Midcareer reassessment A reevaluation of one's career at midlife, which may lead to a career change.

Middle childhood Ages 7 to 11.

Middle ear The part of the ear between the ear drum (part of the outer ear) and the cochlea (part of the inner ear) containing three small bones (ossicles) that efficiently transmit vibration of the ear drum to the cochlea for auditory processing.

Middle phase of labor That phase of labor in which contractions increase in strength and frequency, and the cervix dilates to at least 5 cm, or 2 inches.

Midlife crisis A popular term for the phenomenon of mental and emotional crisis which many people experience between the ages of 40 and 45 years at the end of early adulthood. This is often reflected in the individual's desire to embark on an affair, leave a relationship or make a dramatic change of career.

Midlife transition A change in lifestyle or career that some people may experience at midlife; also, a stage in the culminating phase of early adulthood, ranging from ages 40 to 45.

Midwife A person whose work is helping women in childbirth.

Mind guarding The tendency for members of a group to protect the leader or other decision-makers from input that might influence the leaders to change their minds or raise questions, therefore disrupting the harmony of the organization.

Mindlessness Responding without thinking about the behaviour and its implications.

Mini max principle Principle for improving behaviour through awareness gradually by Minimizing the undesirable behaviour and maximizing the desirable behaviour.

Minimal-sufficiency principle A principle of socialization that holds that children can best internalize ways of acting deemed appropriate if influenced just enough to act as desired, but not enough such that they feel forced to comply.

Minimap In environmental cognition, the information children store about local routes in their environment.

Minimaxing In chess, the technique for mechanically choosing by computer a move in a game.

Minimization A questioning technique in which the interrogator plays down the evidence and the seriousness of the act, by providing an excuse for the act or shifting blame onto someone else (i.e. the victim). See 'maximization'.

Minimum competency tests Criterion-referenced tests focusing on important skills students are expected to have mastered to qualify for promotion or graduation.

Minnesota multiphasic personality inventory (MMPI) An objective test originally designed to distinguish individuals with different psychological problems from normal individuals. It has since become popular as a means of attempting to identify personality characteristics of people in many everyday settings.

Minority group An ethnic or racial group that is a minority within a broader society.

Minority influence The process by which dissenters produce change within a group.

Mirror-image perceptions Reciprocal views of one another often held by parties in conflict; for example, each may view itself as moral and peace-loving and the other as evil and aggressive.

Misattribution of arousal A situation in which the explanation of the physiological symptoms of arousal is switched from the real source to another one.

Miscarriage The natural expulsion of an embryo or foetus from the womb before it is sufficiently developed to survive.

Misinformation effect 1. Incorporating "misinformation" into one's memory of the event, after witnessing an event and receiving misleading information about it.

2. Witnessing an event, receiving misleading information about it, and then incorporating the "misinformation" into one's memory of the event.

Mismatching the way In which a young child will use words wrongly when it does not fully comprehend the adult meaning of them.

Misophobia or mysophobia Fear of being contaminated with dirt or germs.

Mission oriented An offender typology. The offender is directed by a self-imposed task without regard for the consequences of ones actions. The offender is often unconcerned about escape or even survival once the offence has taken place.

Mitigating information Information about a person's situation indicating that he or she should not be held fully responsible for aggressive actions.

Mitochondria Any of various very small, usually rodlike, structures found in the cytoplasm of eukaryotic cells and that serve as a center of intracellular enzyme activity that produces the ATP needed to power the cell.

Mixed crime scene Description for a crime scene the demonstrates the presence of both an organized and disorganized offender. Can be caused by: multiple offenders, unanticipated events, youthfulness, substance abuse, and unexpected victim actions.

Mnemonic devices Various devices that help, or are meant to help, the memory.

Mnemonic system A special technique or strategy consciously employed in an attempt to improve memory.

Mnemonics Strategies or devices that use familiar information during the encoding of new information to enhance subsequent access to the information in memory.

Mnemophobia Fear of memories.

Mock participation Situation in which students appear to be on task but are not engaged with learning.

Mode A measure of central tendency that uses the most frequently occurring score. A distribution with two or more scores that are equal and occur most frequently is called multi-modal.

Model A relatively simple system that works on known principles and is able to do at least some of the things that a more complex system can do.

Modeling In psychotherapy, a technique in which the therapist shows the patient some effective ways of handling problematic situations.

Modern Racism A form of prejudice that surfaces in subtle ways when it is safe, socially acceptable, and easy to rationalize.

Modesty Underrepresenting one's positive traits, contributions, or accomplishments.

Modulation An attempt to exaggerate or minimize the expression of an emotion.

Modus operandi (MO) The offenders actions during an offence. This is variable behaviour that evolves over multiple acts due to offender sophistication and confident.

Modus ponens One of the rules of inference, which states that, no matter what statements p and q stand for, if p is true and if it is also true that if p then q, q must be true as well.

Molysmophobia or molysomophobia Fear of dirt or contamination.

Monists In philosophical discussion of the nature of the human mind, those who believe that mental and physical behaviour can both be explained in common terms, in opposition to dualism, which views mind and matter as being two separate entities.

Monoamine oxidase (MAO) inhibitors The first class of antidepressant medications, such as Nardil and Parnate. They are effective, especially for some kinds of depressions, but they impose difficult dietary restrictions on patients and have been largely superseded by the selective serotonin reuptake inhibitors (ssris).

Monoamine transmitter One of several families of chemicals manufactured by the neurones, such as hormones.

Monochromatic light Light of a single wavelength.

Monocular cues Depth perception cues that require only one eye.

Monocular depth cues Features of the visual stimulus (e.g., linear perspective and motion parallax) that indicate depth even when it is viewed with one eye.

Monogamy A mating pattern in which a reproductive partnership is based on a special, more or less permanent tie between one male and one female.

Monogenic traits Traits determined by a single gene.

Monopathophobia Fear of definite disease.

Monophobia Fear of solitude or being alone.

Mood disorder A disorder characterized by significant shifts or disturbances in mood that affect normal perception, thought, and behaviour. Mood disorders may be characterized by deep, foreboding depression, or a combination of the depression and euphoria.

Moral development and judgment The ability to reason about right and wrong.

Moral dilemmas Hypothetical situations that require a person to consider values of right and wrong.

Moral exclusion The perception of certain individuals or groups as outside the boundary within which one applies moral values and rules of fairness. Moral inclusion is regarding others as within one's circle of moral concern.

Moral realism The first stage of Piaget's model of moral development, which includes egocentrism and blind adherence to rules.

Moral reasoning The reasons and processes that cause people to think the way they do about right and wrong.

Morality A system of beliefs and values that ensures that individuals will keep their obligations to others in society and will behave in ways that do not interfere with the rights and interests of others.

Morality of cooperation The second stage of Piaget's model of moral development, which involves the recognition of rules as social conventions.

Moratorium A suspension of activity of one kind or another; erik erikson has argued that adolescence can be thought of as a period of 'psychosocial moratorium', during which society allows or sanctions young people to try out different beliefs and practices - sexual, religious, political - without undue pressure.

Mores Formal rules of acceptable behaviour that are considered conducive to the welfare of society and so, through general observance, develop the force of law, often becoming part of the formal legal code.

Morning sickness A condition of nausea and vomiting, sometimes accompanied by dizziness, headache, and so on, that affects many women during the first months of pregnancy; it occurs most often in the morning.

Morpheme The smallest significant unit of meaning in a language (e.g., the word *boys* has two morphemes, *boy* and *s*).

Morphic resonance Concept that animals having similar body have similar minds which influence each other psychically by the principle of resonance.

Morphological sex Classification as male or female based on one's sex organs and bodily appearance (e.g., ovaries, vagina, and smooth facial skin versus testes, penis, and facial hair).

Mortality Subject drop-out in a research study. Mortality becomes a problem when a disproportionate drop out rate occurs between two or more groups (Example: 30% of males drop out of group one while only 2% of males drop out in group two, resulting in uneven groups).

Moses Based on work by Sellin, Frazier, and others, Freud speculated two Moses: an Egyptian nobleman who lived near the time of the Egyptian pharaoh Akhenaton, the founder of the first monotheism, and gave the Hebrews a modified version of that religion and was killed by them as a result; and a Midianite who resurrected this religion, modified it still further, and coupled it with the Egyptian rite of circumcision, thereby setting the Hebrews apart as a chosen people.

Motherese A whimsical term for the singsong speech pattern that mothers and other adults generally employ when talking to infants.

Motion after effect A temporary after effect produced by adaption to one pattern of motion, in which stationary stimuli appear to have the opposite pattern of motion.

Motion detectors Cells in the visual cortex that are sensitive to an image moving across the retina.

Motion parallax A depth cue provided by the fact that, as an observer moves, the images cast by nearby objects move more rapidly on the retina than the images cast by objects farther away.

Motion stream The neural pathway in the visual system thought to be concerned primarily with the analysis of stimulus movement.

Motion-sensing The direct measurement of retinal motion by specialized neurones which is thought to underpin the human short range motion system.

Motivated forgetting The idea that people forget things they don't want to remember; also called *psychogenic amnesia.*

Motivation The process of starting, directing, and maintaining physical and psychological activities; includes preferences for one activity over another and persistence of responses.

Motivation to comply In attitude functions, the desire to comply with a particular person or group which affects normative beliefs.

Motive Internal states that provide direction for one's behaviours.

Motoneurons Neurons whose cell bodies are in the brain or spinal cord and whose axons terminate on muscle fibres.

Motor association cortex Those regions of cerebral cortex that control the primary motor cortex; involved in planning and executing behaviours.

Motor development The increasing coordination of muscles that makes physical movements possible.

Motor neuron A neuron whose terminal buttons form synapses with muscle fibres. When an action potential travels down its axon, the associated muscle fibres will twitch.

Motor skills The ability to move with intention.

Motorphobia Fear of automobiles.

Mottephobia Fear of moths.

Mourning The actions or feelings of one who feels or expresses sorrow; specifically, the expression of grief at someone's death.

Movement In theoretical linguistics, where a phrase seems to belong somewhere in a sentence, but can appear somewhere else, such as a topicalization movement.

MRI (magnetic resonance imaging) A noninvasive neurodiagnostic technique that relies on nuclear magnetic resonance. An MRI scan passes a high frequency alternating magnetic field through the brain to detect the different resonant frequencies of its nuclei. A computer then assembles this information to form a picture of brain structure.

Muller-Lyer illusion Illusion in which two lines of the same length appear to be different lengths because of different diagonal lines attached to the end of each line.

Multi infarct dementia Dementia resulting from a stroke.

Multicultural education Education that teaches the value of cultural diversity.

Multicultural research Research designed to examine racial and ethnic groups within cultures.

Multifactor aptitude battery Test that predicts ability to learn a variety of specific skills and types of knowledge.

Multistage sampling A combination of these different methods.

Multi-trait personality inventory In psychological analysis, a quick overall summary of individuality prepared in a test format, enabling comparisons to be drawn across a whole sample.

Multiple-act criterion In the study of personality, a pooled combination of many behaviours deemed to be related to a particular trait or characteristic.

Multiple choice item Test item usually consisting of a stem followed by choices, or alternatives.

Multiple correlation A correlational technique used when there is one X and two or more Y. (Example: the correlation between age and (math and English ability).

Multiple intelligences In Howard Gardner's theory, the six essential, independent mental capacities, some of which are outside the traditional academic notions of intelligence, i.e., linguistic, logical-mathematical, spatial, musical, bodily-kinesthetic, and personal intelligence.

Multiple personality Dissociative mind. Found commonly in primitives. At different times different personalities surface and the person behaves differently. Often the person has amnesia regarding what he did or experienced when he was having another personality.

Multiple personality disorder (MPD) A dissociative disorder in which different aspects of a personality function independently of one another, creating the appearance of two or more distinct personalities within an individual.

Multiple pregnancy A pregnancy in which two or more eggs are fertilized, or a single fertilized egg divides into two or more zygotes.

Multiple sclerosis (MS) A progressive neurological disease wherein the immune system mistakenly destroys the myelin sheaths that comprise the brain's white matter, producing manifestations such as numbness, blindness, and paralysis.

Multiple treatment interference That the treatments interact with each other, thus producing false or misleading results.

Multiplicative rule The probability of the joint occurrence of two or more independent eventis is the product of their individual probabilities.

Mundane realism Degree to which an experiment is superficially similar to everyday situations.

Mundane realism The degree to which the experimental situation resembles places and events in the real world.

Muscle spindle A muscle fibre that functions as a stretch receptor, arranged parallel to the muscle fibres responsible for contraction of the muscle, thus detecting muscle length.

Musophobia or muriphobia Fear of mice.

Mutation In the study of evolution, the mechanism of variation which occurs when the genetic code along the dna molecule in a chromosome is replicated inaccurately, a process which is not always destructive.

Mutual constitution The reciprocal way in which an individual is shaped by the surrounding culture and simultaneously shapes the culture with his or her behaviour. The two modes of mutual constitution are independent (focus on the uniqueness of the individual, being unique) and interdependent (focus on group or community, sense of connection, and responsibility to larger group).

Mutuality synchronous (back-and-forth) Interaction between individuals.

Mutually exclusive Events that occur independently of each other and as a result of chance.

Mycophobia Fear or aversion to mushrooms.

Mycrophobia Fear of small things.

Myctophobia Fear of darkness.

Myelin A fat-rich substance from which the sheaths enclosing axons in the central nervous system are made.

Myelin sheath The series of fatty wrappers, formed by special glial cells, that surround the axons of those neurons that must communicate over long distances in the nervous system and that allow for fast propagation of action potentials along those axons.

Myelination The process in which the brain's neural fibres are coated with an insulating fatty sheath (called myelin), facilitating quicker, more efficient transmission of messages.

Myrmecophobia Fear of ants.

Mythophobia Fear of myths or stories or false statements.

Myxophobia Fear of slime. (Blennophobia).

N

N Symbol used for the number of subjects or data in a distribution. A study with 10 subjects would have an N equal to 10.

Naloxone A drug that blocks the pain alleviation ascribed to endorphins and inhibits the effect of morphine and similar opiates by binding to opiate receptors in the brain.

Name calling A strategy of labeling people in order to influence their or others' thinking.

Narcissism The investment of libido into oneself. The libidinal equivalent of egotism. Freud regarded it as a perversion until 1914, when in "On Narcissism" he wrote that normal development means transferring more and more attention and interest into other people and thereby decreasing one's original or primary state of narcissism. Primary narcissism is the self-involvement all infants start out with; secondary narcissism is a turning of libido away from objects back to the ego, as with what we now call the narcissistic personality.

Narcissistic personality disorder A disorder in which a person has an exaggerated sense of importance, a strong desire to be admired, and a lack of empathy.

Narcolepsy A sleep disorder characterized by an irresistible compulsion to sleep during the daytime.

Narcotics Drugs that can relieve pain; also called *opiates*.

Narrative A mnemonic system in which items to be remembered are linked together by a story.

Narrative method The process of making up a story in order to remember something.

Nativism The view that some important aspects of perception and of other cognitive processes are innate.

Natural concept A category of objects or situations based on people's perceptions and interactions with things in the world; based on exemplars.

Natural selection The explanatory principle by which Darwin accounted for biological evolution. It refers to the greater number of offspring reaching sexual maturity shown by individual organisms possessing hereditary attributes that are advantageous in a given environment. Continued natural selection over many generations can result in wholesale changes in bodily form and behaviour that may result in the development of new species.

Naturalist fallacy The error of defining what is good in terms of what is observable. For example: what's typical is normal; what's normal is good.

Naturalistic observation A method of collecting information about subjects in a natural setting without interfering with them in any way.

Nature vs. Nurture Debate The persistent controversy about whether behaviour or other human characteristics are genetically predetermined, or if they are shaped predominantly by the environment and events in an individual's life.

NCE Score A standard score that sets the mean to fifty and standard deviation to 21.06, allowing the 99th percentile to have a score of 99 and the first percentile a score of 1.

Nebulaphobia Fear of fog. (Homichlophobia).

Necrophilia Sexual gratification from intercourse with a deceased individual. SEE 'insertional necrophelia'.

Necrophobia Fear of death or dead things.

Need for achievement (n Ach) An assumed basic human need to strive for achievement of goals that motivates a wide range of behaviour and thinking.

Need for affiliation The desire to establish and maintain many rewarding interpersonal relationships.

Need for closure The desire to reduce cognitive uncertainty, which heightens the importance of first impressions.

Need for cognition In the study of attitude change, the tendency to engage in and enjoy thinking, which makes people more likely to elaborate the arguments contained in a message.

Need to belong A motivation to bond with others in relationships that provide ongoing, positive interactions.

Negative afterimage In colour vision, the persistence of an image that possesses the hue complementary to that of the stimulus (e.g., seeing a yellow afterimage after staring at a blue lamp), resulting from the operation of opponent processes.

Negative cognitive schema For Aaron Beck, the core cognitive component of depression, consisting of an individual's automatic negative interpretations concerning himself, his future, and the world.

Negative correlation A correlation where one two variables tend to move in the opposite direction (example: the number of pages printed and the amount of ink left in your printer are negatively correlated. The more pages printed, the less ink you have left.).

Negative feedback A process whereby the effect produced by an action serves to diminish or terminate that action. Regulatory systems are characterized by negative feedback loops.

Negative punishment Procedure of removing an event or thing from a situation or preventing a future event from occurring—contingent on the occurrence of a behaviour. Probability of occurrence (POO) of the related behaviour decreases.

Negative reinforcement A consequence that increases the frequency of a response that is regularly and reliably followed by the termination of an aversive stimulus.

Negative reinforcer Release from an unpleasant situation to strengthen behaviour.

Negative skew A frequency curve in which the mean is lower than the median, and the median is lower than the mode. The result is a "tail" towards the more negative side of the graph.

Negative state relief model A theory suggesting that for those in a bad mood, helping others may be a way to lift their own spirits if the perceived benefits for helping are high and the costs are low.

Negative symptoms Indicated by an absence or reduction of normal behaviour.

Negative symptoms of schizophrenia Symptoms that involve deficits in normal functioning, such as apathy, impoverished speech, and emotional blunting.

Negatively skewed distribution A data distribution with a few very low scores.

Negativity bias The tendency for negative traits to be weighted more heavily than positive traits in impression formation.

Neglect syndrome The result of certain lesions of the right parietal lobe that leave a patient inattentive to stimuli to her left (e.g., not eating food on the left side of the plate) and result in her ignoring the left side of her body (e.g., putting makeup on only the right side of her face).

Nelophobia Fear of glass.

NEO Personality Inventory A test that measures the Big Five traits: extraversion, openness to experience, agreeableness, conscientiousness, and neuroticism.

Neocortex The outermost, convoluted layer of the forebrain, often referred to merely as the cortex.

Neo-Freudians A group of theorists who accept the psychoanalytic conception of unconscious conflicts but who differ with Freud in ways that can include 1. Describing these conflicts in social terms rather than in terms of bodily pleasures or frustrations,

2. Maintaining that many of these conflicts arise from specific cultural conditions instead of being biologically preordained.

Neonatal period The first 4 weeks of life outside the womb.

Neopharmaphobia Fear of new drugs.

Neophobia Fear of anything new.

Nephophobia Fear of clouds.

Nerve A bundle of nerve fibres that transmit information between the central nervous system and the body's sense organs, muscles, and glands.

Nerve growth factors Neurochemicals that promote the sprouting of new neuronal connections.

Nerve impulse An electrical discharge passed along a neuron's internal fibre.

Nervous system A complex, highly coordinated network of tissues that communicate via electrochemical signals.

Network model Theories of cognitive organization, especially of semantic memory, which hold that items of information are represented by a system of nodes linked through associative connections.

Neural hierarchy The idea (now unpopular) that a descriptive hierarchy of the stimulus can be achieved simply by wiring together cells with progressively more complex properties. Neural network a type of computer model consisting of many simple units (thought to be roughly equivalent to individual neurones), connected together by excitatory or inhibitory links.

Neural network A type of computer model consisting of many simple units (thought to be roughly equivalent to individual neurones), connected together by excitatory or inhibitory links.

Neural plasticity The capacity for neurons to alter their functioning as a result of experience.

Neural plate A small thickening, running the length of the embryo, from which the neural tube and, eventually, the nervous system, develop.

Neural tube In the early stages of the development of the brain in the embryo, a tubular structure derived from the outer layer of cells of the embryo and from which the brain and spinal cord develop.

Neuro linguistic programming (NLP) A field of human endeavor originally concerned with empirical study of, and modeling of, human performance and excellence, with the goal of creating transferable skill sets, and this remains the core activity of the field to this day.

Neurodevelopmental disorder A disorder that stems from early brain abnormalities. Many researchers believe that schizophrenia is one such disorder and may originate in abnormal foetal brain development.

Neuroethologist A psychologist who specializes in applying the methodologies of brain science to the study of animal behaviour.

Neuroglia The 'nerve glue' which holds neurones together.

Neuroimaging instruments Electronic devices that permit noninvasive study and depiction of brain structure or function.

Neuromodulator A chemical that alters the general sensitivity and responsivity of neurones.

Neuron A nerve cell; consists of a cell body with dendrites and an axon whose branches end in terminal buttons that synapse with muscle fibres, gland cells, or other neurons.

Neuropathic pain Pain caused by abnormal functioning or overactivity of nerves; it results from injury or disease of nerves.

Neuropeptide Y (NPY) A chemical found widely in the brain and periphery. In the brain, it acts as a neurotransmitter; when administered at sites in and near the hypothalamus, it is a potent elicitor of eating.

Neuropsychological assessment A specialized kind of psychological testing used to pinpoint the pattern of cognitive strengths and impairments that occurs with learning disabilities, aging, brain injuries, or diseases.

Neuroscience A field of study which deals with the structure, development, function, chemistry, pharmacology and pathology of the central or peripheral nervous system. The biological study of the brain is an interdisciplinary field, which involves many levels of study, from the molecular level through the cellular level (individual neurons), the level of relatively small assemblies of neurons like cortical columns, that of larger subystems like that which subserves visual perception, up to large systems like the cerebral cortex or the cerebellum, and at the highest level the nervous system as a whole. At this highest level the field largely merges with cognitive neuroscience, a discipline first populated mostly by cognitive psychologists, currently becoming a dynamic specialty of its own. Thus, the concern of neuroscience includes such diverse topics as the operation of neurotransmitters, how genes contribute to the embryonic development of the nervous system and to learning, the operation of relatively simpl.

Neurosis A conflict between ego and id that produces symptoms of psychological discomfort. This physiologic-sounding term has been changed to "anxiety disorder." That's regrettable, since many neuroses don't involve conscious anxiety.

Neurosis results because we live in cultures that require for their cohesion a renunciation of our drives, a repression that makes us a bit irrational.

Neurotic disorders Mental disorders in which a person does not have signs of brain abnormalities and does not display grossly irrational thinking or violate basic norms but does experience subjective distress; a category dropped from DSM-III.

Neuroticism A trait dimension that refers to emotional instability and maladjustment.

Neuroticism, extraversion, and openness personality inventory (NEO-PI) The instrument used to measure the elements described in the five-factor model of personality (neuroticism, extraversion, openness, agreeableness, and conscientio-usness).

Neurotoxin Any chemical poisonous to neurons.

Neurotransmitter Neurotransmitters are chemicals that are used to relay, amplify and modulate electrical signals between two neurons: the presynaptic neuron and the postsynaptic neuron.

Neutral stimuli Stimuli that do not naturally prompt a particular response.

Niches In the study of ecology, the environmental slot containing all the determinants of survival into which a particular species fits after going through a period of adaptation.

Nightmare A frightening dream occurring in REM sleep.

Nirvana principle Barbara Low's term for the psychological equivalent of homeostasis, the push for the least amount of tension. Different from the pleasure principle in that
1. pleasure sometimes increases with tension, and
2. the Nirvana Principle is primarily under the sway of the death drive, whereas the pleasure principle is powered by Eros.

Nociceptive pain Pain induced by a noxious external stimulus; specialized nerve endings in the skin send this pain message from the skin, through the spinal chord, into the brain.

Noctiphobia Fear of the night.

Nocturnal emissions Signal of the onset of puberty for boys; also called *wet dreams*.

Node In environmental cognition, a place where routes intersect in the mental images people develop of maps.

Nodes of Ranvier The gaps occurring between the glial-cell wrappers that form the myelin sheath surrounding many kinds of axons. The nodes are crucial to the rapidity at which neural impulses travel along myelinated axons.

Nomatophobia Fear of names.

Nominal fallacy The false belief that one has explained the causes of a phenomenon by identifying and naming it; for example, believing that one has explained lazy behaviour by attributing it to "laziness.".

Nominal measurement "naming:" assigns arbitrary numbers to different categories in order to differentiate them. The numbers themselves have no meaning. For example, you may bubble in the number "1" if you're a female, "2" if you're a male. This doesn't mean that either category is better or worse than the other.

Nominal scale Any scale that contains no magnitude. Often nominal is thought of as name only, meaning that the variables of a nominal scale can be identified but not measured.

Non-accidental property A property of the image that probably reflects the same property of the world, such as a straight line in an image which is almost (but not quite) certainly produced by a straight edge.

Non-common effects In the process of determining a dispositional attribution, the number of very specific effects caused by an action.

Non directional A hypothesis which predicts that there will be a difference between or among groups.

Non equivalent control group design Designs with experimental and control groups that are designated before the treatment occurs and are not created by random assignment.

Non experimental design A design in which no treatments are given, instead studying naturally occurring variation in the independent and dependent variables. Used when an experimental design would be unethical.

Nonconscious Information not typically available to consciousness or memory.

Nondeclarative memory A subsystem within Long term memory which consists of skills we acquire through repetition and practice (e.g., dance, playing the piano, driving a car).

Nondirective techniques A set of psychotherapy techniques devised by Carl Rogers. As far as possible, the counselor refrains from offering advice or interpretation but only tries to clarify the patient's own feelings by echoing or restating what he says.

Nonfluent aphasia Speech disorder in which the main difficulty is in speech production, often involving damage to Broca's area in the frontal lobe.

Nongraded programs (cross-age grouping programs) Programs that combine children of different ages in the same class, generally at the primary level.

Nonparametric test Any statistic that is designed for ordinal or nominal data or data that is not normally distributed.

Non-REM (NREM) sleep The period during which a sleeper does not show rapid eye movement; characterized by less dream activity than REM sleep.

Nonsense syllable Two consonants with a vowel between that do not form a word. Used to study associations between relatively meaningless items.

Nonverbal behaviour Communicating feelings and intentions without words.

Nonverbal behaviour or nonverbal communication Behaviour that reveals a person's feelings without words—through facial expressions, body language, and vocal cues.

Non-verbal communication A form of communication without speech developed between babies and their carers, which makes use of faces, voices, body movements and touch, but which adults and the disabled continue to use throughout life to supplement and to support speech or act as its substitute.

Nonverbal cues Eye contact, gestures, physical proximity, or touching used to communicate without interrupting verbal discourse.

Non-verbal vocal tone In channels of communication, the reinforcement of words by gesture.

Nonviable foetus A foetus that is unable to live on its own.

Non-zero-sum games Games in which outcomes need not sum to zero. With cooperation, both can win; with competition, both can lose. (also called mixed-motive situations.).

Norepinephrine (NE) The neurotransmitter found in the nerves of the sympathetic branch of the ANS. It is also one of the neurotransmitters involved in various arousal systems in the brain.

Norm An expected standard of behaviour and belief established and enforced by a group.

Norm crystallization The convergence of the expectations of a group of individuals into a common perspective as they talk and carry out activities together.

Norm of social justice A social norm stating that we should help only when we believe that others deserve our assistance.

Norm of social responsibility A moral standard emphasizing that people should help those who need assistance.

Norm referenced evaluations Assessments that compare the performance of one student against the performance of others.

Normal curve The normal distribution, also called Gaussian distribution, is an extremely important probability distribution in many fields, especially in physics and engineering. It is a family of distributions of the same general form, differing in their location and scale parameters: the mean ("average") and standard deviation ("variability"), respectively. The standard normal distribution is the normal distribution with a mean of zero and a standard deviation of one (the green curves in the plots to the right). It is often called the bell curve because the graph of its probability density resembles a bell.

Normative beliefs In attitude functions, those beliefs about whether specific others would approve of the person performing the behaviour which determine the subjective norm component.

Normative development A typical pattern of development common to many children used as the basis for descriptive work by psychologists.

Normative influence Conformity based on a person's desire to fulfill others' expectations, often to gain acceptance.

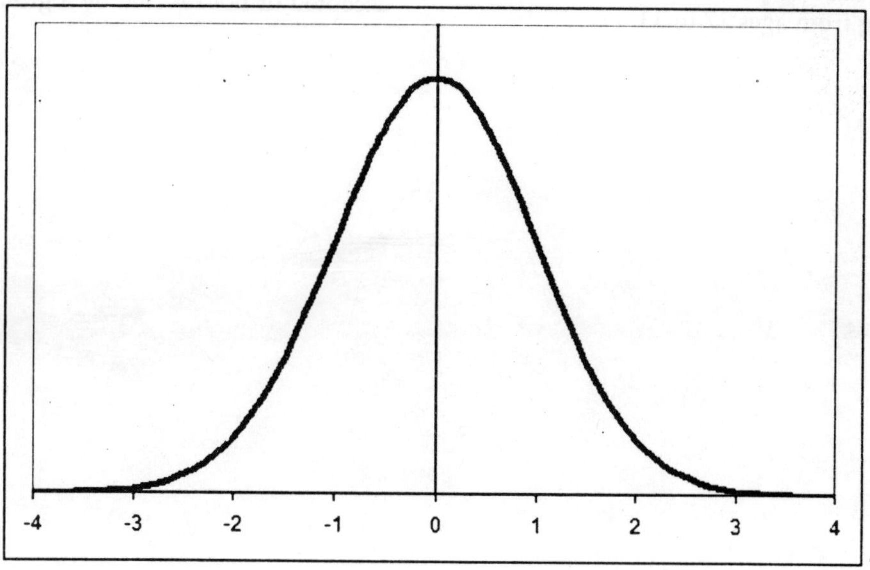

Fig. Normal curve

Normal curve equivalent Set of standardized scores ranging from 1 to 99, having a mean of 50 and a standard deviation of about 21.

Normal distribution A frequency distribution whose graphic representation has a symmetric, bell-shaped form — the normal curve. Its characteristics are often referred to when investigators test statistical hypotheses and make inferences about the population from a given sample.

Normative investigations Research efforts designed to describe what is characteristic of a specific age or developmental stage.

Normative model of leadership The theory that leadership effectiveness is determined by the amount of feedback and participation that leaders invite from workers.

Nyctophobia Fear of the dark or of night.

Nyctohylophobia Fear of dark wooded areas or of forests at night.

Numerophobia Fear of numbers.

Null hypothesis The assumption that there is no true difference between groups and any difference (statistically) is due to sampling errors. A researcher tries to disprove this.

Nudophobia Fear of nudity.

Nucleus accumbens A dopamine-rich area in the forebrain that is critical in the physiology of reward.

Nucleotides Biochemical units that make up DNA and genes.

Nucleomituphobia Fear of nuclear weapons.

Novice phase A phase of early adulthood, ranging from ages 17 to 33.

Novercaphobia Fear of your step-mother.

Novelty seeking Tendency to seek out-of-the-ordinary experiences.

Note taking A study strategy that requires decisions about what to write.

Nostophobia Fear of returning home.

Nosophobia or nosemaphobia Fear of becoming ill.

Nosocomephobia Fear of hospitals.

Normative social influence An individual's tendency to conform because of a need to be accepted or not rejected by a group.

O

Obedience Social influence in which one person obeys direct orders from another to perform some action. High obedience levels were seen in nazi germany, in cults, and often in the childhood of psychopathic offenders.

Obesity A condition of marked overweight in animals and humans produced by a large variety of factors including genetic predisposition ("thrifty" genes), metabolic factors (oversecretion of insulin), and behavioural conditions (overeating, insufficient exercise).

Obesophobia Fear of gaining weight. (Pocrescophobia).

Object A mental representation of a person. Also called an imago.

Object permanence The idea posited by jean piaget that objects which are no longer visible still continue to exist even though they are out of sight, which although taken for granted by adults requires the forming of a mental representation on the part of the infant.

Object relations The relationships that people have with others, who are represented mentally as objects with certain attributes.

Object relations theory Psychoanalytic theory that originated with Melanie Klein's view that the building blocks of how people experience the world emerge from their relations to loved and hated objects (significant people in their lives).

Objective personality tests Tests that usually consist of self-report inventories. Commonly used objective tests include the MMPI-2, the 16PF, and the NEO Personality Inventory.

Objective Techniques A generic term for the psychological procedures used to measure personality which rely on measurable or objective techniques such as the MMPI-2 and WAIS-III.

Objective tests Objective tests are different from obtrusive tests, because objective tests are not projective in nature. Projective tests are really Freudian and teach people about their fantasies and objectives are true and false self-reports about themselves. These are the main differences.

Observational approach A non-experimental case study of human behaviour which involves the keeping of an accurate record of events which can be studied in detail later.

Observational learning A mechanism of socialization whereby a child observes another person who serves as a model and then proceeds to imitate what that model does.

Observational research Research based on the observation of subjects in laboratory or natural settings rather than on experimentation or interviews.

Observational study A study in which the investigator does not manipulate any of the variables but simply observes their relationship as they occur naturally.

Observer bias The distortion of evidence because of the personal motives and expectations of the viewer.

Obsession An involuntary recurring thought, idea, or image.

Obsessive compulsive Feeling compelled to think for some apparently meaningless idea or do an apparently meaningless act.

Obsessive-compulsive disorder A disorder whose symptoms are obsessions (persistent and irrational thoughts or wishes) and compulsions (uncontrollable, repetitive acts), which seem to be defences against anxiety. A member of a diagnostic category called anxiety disorders, which also includes generalized anxiety disorder and phobias.

Occipital lobe One of for lobes of the brain. Contains the visual cortex and therefore plays a major role in the interpretation of visual information.

Occlusion The partial concealment of one object by another object in front of it.

Occupational psychology The study of work in terms of the feelings, attitudes, behaviour and performance of those at work and those factors which influence workers, attitudes and behaviour.

Ochlophobia Fear of crowds or mobs.

Ochophobia Fear of vehicles.

Octophobia Fear of the figure 8.

Odd even reliability The correlation coefficient determined by comparing odd items of the measurement to the even items. One method to determine the internal consistency of a test or measuring device.

Odontophobia Fear of teeth or dental surgery.

Odynophobia or odynephobia Fear of pain. (Algophobia).

Oedipal complex The boy's tendency, around the age of five, to experience his freshly awakened sexual strivings toward his mother while wanting to replace his father in her affections. Mostly unconscious. When successfully resolved, these feelings are fully repressed, and the boy, afraid of castration, learns to identify with his father. As a result of all this, he internalizes his parents and acquires a superego whose ego ideal replaces some of his early narcissism. The feminine equivalent has been named the Electra Complex, but for Freud, women have inferior superego development and therefore an inferior conscience because they never have to disidentify with mother as boys do.

Freud saw an unresolved Oedipal Complex as the cause of every significant neurosis. The name comes from King Oedipus, who killed his father, married his mother unknowingly, and put his own eyes out (which Freud interprets as symbolic castration) when he discovered the truth of his origins. In a late innovation, Freud was prompted by the work of female analysts like Helene Deutsch to emphasize the initial pre-Oedipal strength of the mother-daughter bond.In Freud's theory, an unconscious psychological conflict in which the child loves the parent of the opposite sex. In psychoanalytic theory, a general term for the cluster of impulses and conflicts that occurs during the phallic phase, at around age five. In boys, a fantasized form of intense, possessive sexual love is directed at the mother, which is soon followed by hatred for and fear of the father. As the fear mounts, the sexual feelings are pushed underground and the boy identifies with the father. An equivalent process in girls is called the Electra complex. The crux of freud's theory of personality, the unconscious sexual desire of a son for his mother and consequent jealousy and hatred of his father. Freud considered this complex to be universal and, if unresolved, to be at the seat of all other neuroses and the cause of a great deal of guilt. Freud was fascinated by the myth of Oedipus, the tragic hero (of Sophocles, the Greek dramatist) who unwittingly killed his father and married his mother. Freud thought that, during the phallic stage, the young boy develops an intense sexual love for his mother. Because of this, he sees his father as a rival, and wants to get rid of him. The father, however, is far bigger and more powerful than the young boy, and so the child develops a fear that, seeing him as a rival, his father will castrate him. Because it is impossible to live with the continual castration-threat anxiety provided by this conflict, the young boy develops a mechanism for coping with it, using a defence mechanism known as 'identification with the aggressor'. He stresses all the ways that he is similar to his father, adopting his father's attitudes, mannerisms and actions, feeling that if his father sees him as similar, he will not feel hostile towards him. A sequence of development experiences that Freud argued all human boys went through. It involves the boys romantic feelings for his mother. However, in Freud's theory, if the oedipal ocmplex is "properly" resolved, the boy gives up his quest for his mother's romantic affections. Supposedly this happens because he believes, unconsciously and symbolically at least, that he will otherwise lose his penis (be castrated.).

Oedipal stage Overlapping some with the phallic stage, this phase (ages 4 to 6) represents a time of inevitable conflict between the child and parents. The child must desexualize the relationship to both parents in order to retain affectionate kinship with both of them. The process is accomplished by the internalization of the images of both parents, thereby giving more definite shape to the child's personality. With this internalization largely completed, the regulation of self-esteem and moral behavior comes from within.

Oedipus conflict The unconscious tendency of a child to be attached to the parent of the opposite sex and hostile toward the other parent. Its persistence in adult life results in neurotic disorders. Originally restricted to a son's attachment.

Oenophobia Fear of wines.

Off-centre cell A type of retinal ganglion cell with a receptive field in which the centre sub-region is an on-region.

Off-region A region of the receptive field of a sensory neurone in which an increase in stimulus intensity produces a decrease in response whilst a decrease in stimulus intensity produces an increase in response.

Oikophobia Fear of home surroundings, house.(Domatophobia, Eicophobia).

Old-fashioned racism Blatantly negative stereotypes based upon White racial superiority, coupled with open opposition to racial equality.

Olfaction The sense of smell.

Olfactophobia Fear of smells.

Olfactory bulb The center where odor-sensitive receptors send their signals, located just below the frontal lobes of the cortex.

Olfactory epithelium The small area at the top of the nasal cavity that contains receptors reactive to airborne chemicals.

Olfactory mucosa The mucous membrane lining the top of the nasal sinuses; contains the cilia of the olfactory receptors.

Oligodendrocyte In the central nervous system, a type of glial cell which produces an insulating sheath around the axons of neurones.

Ombrophobia Fear of rain or of being rained on.

Ommetaphobia or ommatophobia Fear of eyes.

On-centre cell A type of retinal ganglion cell with a receptive field in which the centre sub-region an on-region.

One tailed The type of test used with a directional hypothesis, using +/- 1.65 as the critical value.

One way ANOVA An Analysis of Variance used when there is only one main effect.

One word stage The stage in a child's language development when the elemental aspects of speech have been mastered, and complete words are used to express relationships between people and objects. In other words, the understanding and use of words as symbols, usually at the end of a baby's first year.

Oneirogmophobia Fear of wet dreams.

Oneirophobia Fear of dreams.

One-trial learning In classical conditioning, the establishment of a conditioned response (CR) after only one pairing of conditioned stimulus (CS) and unconditioned stimulus (US).

Only child A child without siblings.

Onomatophobia Fear of hearing a certain word or of names.

On-region A region of the receptive field of a sensory neurone in which an increase in stimulus intensity produces an increase in response, whilst a decrease in stimulus intensity produces a decrease in response.

Ontogenesis In environmental psychology, the development from infancy of cognitive skills for environmental understanding.

Open-ended interview A face-to-face set of questions posed to a subject in the course of social research.

Openness to Experience Tendency to be intellectual, interested in the arts, emotionally aware, and liberal.

Operant Behaviour emitted by an organism that can be characterized in terms of the observable effects it has on the environment.

Operant behaviour/Conditioning A variation on instrumental conditioning in which behaviour operates upon the environment and produces consequences, and conditioning manifests as the change that takes place when those consequences have a particular effect.

Operant chamber An apparatus in which an animal's behaviour can be easily observed, manipulated, and automatically recorded. Sometimes called a "Skinner Box.".

Operant conditioning A form of learning in which behaviour is affected by its consequences. Favourable consequences strengthen the behaviour and unfavourable consequences weaken the behaviour.

Operant extinction When a behaviour no longer produces predictable consequences, its return to the level of occurrence it had before operant conditioning.

Operational definition A definition of a variable or condition in terms of the specific operation or procedure used to determine its presence.

Ophidiophobia Fear of snakes. (Snakephobia).

Ophthalmophobia Fear of being stared at.

Opioid A neuromodulator whose action is mimicked by a natural or synthetic opiate, such as opium, morphine, or heroin.

Opiophobia Fear medical doctors experience of prescribing needed pain medications for patients.

Opponent-process theory of motivation A theory that asserts that the nervous system tends to counteract any deviation from the neutral point on the pain-pleasure dimension. If the original stimulus is maintained, there is an attenuation of the emotional state one is in; if it is withdrawn, the opponent process reveals itself, and the emotional state swings sharply in the opposite direction.

Optic array The complex but structured pattern of light rays that arrives at any given point in the world and which is, in principle, the visual stimulus available to an observer at that point.

Optic chiasm A junction in front of the hypothalamus and the pituitary gland where about half the nerve fibres from each eye cross to the other side of the brain optic flow the smooth changes in the optic array produced as an observer moves about the world.

Optic disk A circular structure located at the exit point from the retina of the axons of the ganglion cells that form the optic nerve.

Optic flow The phenomenon wherein an object's retinal image enlarges as we approach the object and shrinks as we retreat from it. It is used as a depth cue by the visual system.

Optic nerve A bundle of ganglion cell axons that originate in the retina.

Optimal level of arousal Theory arguing that humans are driven to increase or decrease arousal to produce a comfortable level that is not over- nor under stimulating.

Optimal level of skill performance According to Fischer's Skill Model, the brain's maximal capacity for information processing.

Optimism The tendency to expect positive outcomes.

Optimistic Expecting the best in this best of all possible worlds.

Optimistic explanatory style A habitual tendency to attribute negative events to external, unstable, and specific causes, and positive events to internal, stable, and global causes.

Optimum-level hypothesis The hypothesis that organisms will perform behaviour that restores the level of arousal to an optimum level.

Optophobia Fear of opening one's eyes.

Oral aggressive personality Stemming from the Oral stage, a child who becomes fixated due to over stimulation transfers his or her unresolved oral issues into aggression and hostility.

Oral character Forever wanting to suck, to consume, to take in, endlessly hungry and needy.

Oral receptive personality Stemming from the Oral stage, a child who becomes fixated due to under stimulation transfers his or her unmet oral needs into smoking, drinking, talking, biting fingernails, or sucking one's own thumb, for example.

Oral stage In psychoanalytic theory, the earliest stage of psychosexual development during which the primary source of bodily pleasure is stimulation of the mouth and lips, as in sucking at the breast.

Orbitofrontal cortex A region of the prefrontal cortex that plays an important role in recognition of situations that produce emotional responses.

Order effects The effects of administering treatments in a particular order.

Ordered array A table consisting of data in order of highest to lowest or lowest to highest where each data is given a numbered rank depicting it's difference from the highest or lowest score.

Ordinal measurement "rank:" orders participants from high to low (or vice versa) but does not indicate how much higher or lower one subject is from another.

Ordinal scale A scale in which responses are rank-ordered by relative magnitude but in which the intervals between successive ranks are not necessarily equal.

Orexins Hormones synthesized in the lateral hypothalamus that are potent elicitors of eating.

Organic amnesia A permanent form of memory loss, resulting from biological devastation to the brain, such as disease, alcoholism, chemical poisoning, and senility.

Organic brain syndrome Mental deterioration, also known as dementia.

Organismic variables The inner determinants of an organism's behaviour.

Organizational effect An effect of a hormone that usually occurs during prenatal development and produces permanent changes that alter the subsequent development of the organism.

Organizational psychologists Psychologists who study various aspects of the human work environment, such as communication among employees, socialization or enculturation of workers, leadership, job satisfaction, stress and burnout, and overall quality of life.

Organized Offender typology characterized by a mobile (own transportation) offender who cons his victim into capture rather then using force. Offender is often a stranger selected on the basis of specific criteria. Use of restraints and weapons are preplanned, and rarely left behind. The body is often transported to a novel dump site and concealed. SEE 'disorganized'.

Orgasm In women, vaginal rather than clitoral, the latter being a symptom of immaturity.

Oriental psychology Psychology based on ancient Oriental wisdom, concepts and methods.

Orientation constancy The ability to perceive the actual orientation of objects in the real world despite their varying orientation in the retinal image.

Orientation selectivity A property of many cells in the visual cortex, where the cell responds to stimuli of one orientation better than to stimuli of other orientations.

Orienting response Any response by which an organism directs appropriate sensory organs (eyes, ears, nose) toward the source of a novel stimulus.

Ornithophobia Fear of birds.

Orthophobia Fear of property.

Oscilloscope An electronic monitoring device that uses a cathode ray tube (CRT) to display electrical signals such as electrocardiograph signals or action potentials.

Osmophobia or osphresiophobia Fear of smells or odors.

Osmoreceptors Receptors that help to control water intake by responding to the concentrations of bodily fluids.

Ossicle One of the three bones of the middle ear (the *hammer, anvil,* and *stirrup*) that transmit acoustical vibrations from the eardrum to the membrane behind the oval window of the cochlea.

Ostraconophobia Fear of shellfish.

Ought self In the perception of self, the view of the self in terms of 'me as i think i should or ought to be'.

Ouranophobia or uranophobia Fear of heaven.

Outcome dependence Dependence upon others for positive outcomes or rewards (also know as normative dependence).

Outcome measures In psychopathology, variables (e.g., mood ratings, work absenteeism) assessed to indicate whether a particular treatment was effective or cost-effective.

Outcomes-based education An approach to instruction and school organization that clearly specifies what students should know and be able to do at the end of a course of study.

Outer ear The portion of the structures of the ear that includes the earflap, the auditory canal, and the outer surface of the eardrum.

Outgroup A social group with which one does not identify or to which one does not belong.

Outgroup homogeneity effect A phenomenon related to stereotyping in which a member of a group (the in-group) tends to view members of another group (the out-group) as more alike (less varied) than are members of his or her own group.

Outlining Representing the main points of material in heirarchical format.

Oval window An opening in the bone surrounding the cochlea. The stirrup presses against a membrane behind the oval window and transmits sound vibrations into the fluid within the cochlea.

Overcompensation According to Alfred Adler, the attempt to cover up a sense of inferiority by focusing on outward signs of superiority such as status, wealth, and power.

Overconfidence effect The tendency for people to be too certain that their beliefs,.

Overconfidence phenomenon The tendency to be more confident than correct to overestimate the accuracy of one's beliefs.

Overextension A type of error incurred when a child seeks to identify something verbally without comprehending the full adult meaning of the word it is using, such as using the word 'dog' to refer to cats as well as dogs.

Overjustification effect The result of bribing people to do what they already like doing; they may then see their action as externally controlled rather than intrinsically appealing.

Overkill Injury and trauma that is excessive beyond that required to cause the death of the victim.

Overlapping A teacher1s ability to respond to behaviour problems without interrupting a classroom lesson.

Overlearning A technique used to improve memory where information is learned to the point that it can be repeated without mistake more than one time.

Overregularization A grammatical error, usually appearing during early language development, in which rules of the language are applied too widely, resulting in incorrect linguistic forms.

Overregularization errors Errors in speech production in which an irregular noun or verb is treated as though it were regular (e.g., *foots* for *feet* or *goed* for *went*). These errors are often made by young children, which suggests that children discover the rules of language and do not simply imitate what they hear.

Overtone The frequencies of complex tones that occur at multiples of the fundamental frequency.

Ovulate To produce and discharge ova from the ovary.

Ovum A mature female germ cell which, only after fertilization, develops into a zygote and then a foetus.

Own-race bias The tendency for people to more accurately recognize faces of their own race.

P

Pacinian corpuscle A specialized, encapsulated somatosensory nerve ending, which detects mechanical stimuli, especially vibrations.

Pagophobia Fear of ice or frost.

Pain The aversive sensation that usually accompanies tissue damage and alerts the organism to engage in protective or reparative behaviour.

Paired associate learning A task involving the linkage of two items in a pair so that when one is presented the other can be recalled.serial learning A task requiring recall of a list of items.

Panic A feeling of fear mixed with hopelessness or helplessness.

Panic attack A sudden episode consisting of terrifying bodily symptoms such as labored breathing, choking, dizziness, tingling in the hands and feet, sweating, trembling, heart palpitations, and chest pain. Panic attacks occur in a number of mental disorders and are common in phobias, panic disorder, and post-traumatic stress disorder (PTSD).

Panic disorder An anxiety disorder in which sufferers experience unexpected, severe panic attacks that begin with a feeling of intense apprehension, fear, or terror.

Panophobia or pantophobia Fear of everything.

Panthophobia Fear of suffering and disease.

Papaphobia Fear of the Pope.

Papillae Small bumps on the skin that hold taste buds, which in turn hold the taste receptors in the tongue and throat, on the inside of the cheeks, and on the roof of the mouth.

Papyrophobia Fear of paper.

Paralipophobia Fear of neglecting duty or responsibility.

Parallel distributed processing (PDP) Models of cognitive processing in which the relevant symbolic representations do not correspond to any one unit of the network but to the state of the network as a whole.

Parallel distributed processing A model based on the idea that information is processed simultaneously in the sensory register, short-term memory, and long-term memory.

Parallel forms Different versions of a test used to assess test reliability; the change of forms reduces effects of direct practice, memory, or the desire of an individual to appear consistent on the same items.

Parallel forms reliability Two tests of different forms that supposedly test the same material will give the same results.

Parallel play Play in which children engage in the same activity side by side but with very little interaction or mutual influence.

Parallel processes Two or more mental processes that are carried out simultaneously.

Parallel search The simultaneous comparison of a target stimulus to several items in memory.

Parameter One of the ways that languages can depart from a hypothesized universal linguistic structure (e.g., omission of the subject in certain sentences).

Paranoia A mental disorder characterized by systematized delusions, as of grandeur or, especially, persecution; often, except in a schizophrenic state, occurring within an otherwise relatively intact personality.

Paranoid schizophrenia A subcategory of schizophrenia. Its dominant symptom is a set of delusions that are often elaborately systematized, usually of grandeur or persecution.

Paranoid type A subtype of schizophrenia characterized by marked delusions or hallucinations and relatively normal cognitive and emotional functioning.

Paraphobia Fear of sexual perversion.

Paraphrase The relation between two sentences whose meanings (underlying structures) are the same but whose surface structures differ (e.g., "The boy hit the ball"/"The ball was hit by the boy").

Parapraxis Freud never used this word. He wrote of the Fehlleistung, the faulty achievement, when we mean to do one thing and do another. Slips of the tongue are one example.

Parapsychology The branch of Psychology studying behaviour which appears to transcend sensory motor limitations.

Parasitophobia Fear of parasites.

Paraskavedekatriaphobia Fear of Friday the 13th.

Parasympathetic branch The portion of the autonomic nervous system that activates functions that occur during a relaxed state.

Parasympathetic division The subdivision of the autonomic nervous system that monitors the routine operation of the body's internal functions and conserves and restores body energy.

Parasympathetic system A division of the autonomic nervous system that serves vegetative functions and conserves bodily energies (e.g., slowing heart rate). Its action is often antagonistic to that of the sympathetic system.

Parental control The degree to which parents are restrictive in their use of parenting techniques.

Parental investment The resources including time, physical effort, and risks to life that a parent spends in procreation and in the feeding, nurturing, and protecting of the resulting offspring.

Parental warmth The degree to which parents are loving, affectionate, and approving in their use of parenting techniques.

Parenting practices Specific parenting behaviours that arise in response to particular parental goals.

Parenting styles The manner in which parents rear their children; an authoritative parenting style, which balances demandingness and responsiveness, is seen as the most effective.

Parietal cortex The part of the brain that determines where objects are in space, and consequently how one needs to react to them.

Parietal lobe One of four lobes of the brain. Contains the Somatosensory Cortex d is therefore involved in the processing of touch, pressure, temperature, and pain.

Parkinson's disease A degenerative neurological disorder characterized by various motor, difficulties that include tremor, muscular rigidity, and slowed movement. This disease involves degeneration of dopamine-releasing neurons in the basal ganglia of the forebrain, which are crucial for motor control.

Parse tree The structure of the phrases captured by linguistic rules (particularly phrase structure grammar rules) which is often illustrated graphically as an inverted tree. Particulate inheritance in the study of heredity, the separate inheritance of characteristics independent of each other.

Parsing The dissection of a complex stimulus into meaningful parts.

Part learning Mastering new material by learning it one part or subskill at a time.

Parthenophobia Fear of virgins or young girls.

Partial reinforcement A condition in which repeated responses are reinforced only some of the time.

Partial reinforcement effect The behavioural principle that states that responses acquired under intermittent reinforcement are more difficult to extinguish than those acquired with continuous reinforcement.

Participant modeling A therapeutic technique in which a therapist demonstrates the desired behaviour and a client is aided, through supportive encouragement, to imitate the modeled behaviour.

Participant observation Research that requires an observer to become a member of his or her subjects' community.

Parturiphobia Fear of childbirth.

Parturition The act of bringing forth young; childbirth.

Parvo cells Ganglion cells found throughout the retina that, because of their sensitivity to differences in hue, are particularly suited to the perception of colour and form.

Passion Intense feelings of physiological arousal and excitement.

Passionate love A state of intense longing for union with another. Passionate lovers are absorbed in one another, feel ecstatic at attaining their partner's love, and are disconsolate on losing it.

Passive euthanasia The deliberate withdrawal or withholding of life-sustaining treatment that may otherwise prolong the life of the dying person.

Pastoral counselor A member of a religious order who specializes in the treatment of psychological disorders, often combining spirituality with practical problem solving.

Pathogen Disease-producing microbes, including viruses, fungi, and bacteria, that can trigger the production of fever. In evolutionary terms, defence against pathogens may be the driving force behind sexual reproduction.

Pathology model As used in the text, the pathology model describes a general conception of mental disorders which holds that 1. one can generally distinguish signs and symptoms from their underlying causes, and

2. These causes may be regarded as a form of disease.

Pathophobia Fear of disease.

Patient The term used by those who take a biomedical approach to the treatment of psychological problems to describe the person being treated.

Patroiophobia Fear of heredity.

Pattern recognition The process by which the perceptual system matches the form of a figure against the figure as represented in memory.

Pattern theory The theory that a stimulus attribute is not coded by being sent along specific sensory fibres, but rather by a specific pattern of firing of all the relevant sensory fibres.

Pavlov, Ivan (1849-1936) A Russian physiologist who won the Nobel prize for his work on the digestive system of dogs. In the course of his work he chanced upon a phenomenon he could not explain and followed it (reluctantly) out of physiology and into psychology. What puzzled him was that his dogs began to salivate not only when they were presented with food but even before they were fed, when they recognised the man coming to feed them. The series of experiments he designed in an attempt to find the causes of this behaviour became known as classical conditioning and Pavlov spent the last thirty years of his life working out the applications and the implications of his discovery. Though Pavlov thought he had found a way of studying the brain, rather than behaviour, his work inspired a new American school of psychology called behaviourism. Ivan Pavlov was born in a small village in central Russia. His family hoped that he would become a priest, and he went to a theological seminary. After reading Charles Darwin, he found that he cared more for scientific pursuits and left the seminary for the University of St. Petersburg. There he studied chemistry and physiology, and he received his doctorate in 1879. He continued his studies and began doing his own research in topics that interested him most: digestion and blood circulation. His work became well known, and he was appointed professor of physiology at the Imperial Medical Academy. The work that made Pavlov a household name in psychology actually began as a study in digestion. He was looking at the digestive process in dogs, especially the interaction between salivation and the action of the stomach. He realized they were closely linked by reflexes in the autonomic nervous system. Without salivation, the stomach didn't get the message to start digesting. Pavlov wanted to see if external stimuli could affect this process, so he rang a metronome at the same time he gave the experimental dogs food. After a while, the dogs – which before only salivated when they saw and ate their food – would begin to salivate when the metronome sounded, even if no food were present. In 1903 Pavlov published his results calling this a "conditioned reflex," different from an innate reflex, such as yanking a hand back from a flame, in that it had to be learned. Pavlov called this learning process (in which the dog's nervous system comes to associate the sound of the metronome with the food, for example) "conditioning." He also found that the conditioned reflex will be repressed if the stimulus proves "wrong" too often. If the metronome sounds repeatedly and no food appears, eventually the dog stops salivating at the sound.

Pavlov was much more interested in physiology than psychology. He looked upon the young science of psychiatry a little dubiously. But he did think that conditioned reflexes could explain the behavior of psychotic people. For example, he suggested, those who withdrew from the world may associate all stimulus with possible injury or threat. His ideas played a large role in the behaviorist theory of psychology, introduced by John Watson around 1913.

Pavlov was held in extremely high regard in his country – both as Russia and the Soviet Union – and around the world. In 1904, he won the Nobel Prize in physiology/ medicine for his research on digestion. He was outspoken and often at odds with the Soviet government later in his life, but his world renown, and work that his nation was proud of, kept him free from persecution. He worked actively in the lab until his death at age 87.

Payoff matrix In a signal detection experiment, a table that shows the costs and benefits of each of the four possible outcomes: a hit, reporting the stimulus when it is present; a correct negative, reporting it as absent when it is in fact absent; a miss, failing to report it when it is present; and a false alarm, reporting it as present when it is not.

Peace psychology An interdisciplinary approach to the prevention of nuclear war and the maintenance of peace.

Peak experience As Maslow considered it, a profound and deeply felt moment in a person's life, sometimes said to be more common in self-actualized persons than in others.

Pearson Product-Moment correlation A correlation statistic used primarily for two sets of data that are of the ratio or interval scale. The most commonly used correlational technique.

Peccatophobia Fear of sinning or imaginary crimes.

Pedagogy The study of teaching and learning with applications to the instructional process.

Pediculophobia Fear of lice.

Pediophobia Fear of dolls.

Pedophilia An abnormal condition in which an adult has a sexual desire for children.

Pedophobia Fear of children.

Peer pressure To be forced or compelled to do something by one's peers.

Peer tutoring One student teaching another.

Peers People who are equal in age or status.

Peg word method A mnemonic system in which items to be remembered are associated with a set of mental pegs that one already has in memory, such as key words of a rhyme.

Peladophobia Fear of bald people.

Pellagrophobia Fear of pellagra.

Peniaphobia Fear of poverty.

Penis Envy A subset of the castration complex, it involves the supposed envy of "castrated" women of the male member, which they later seek to possess via pregnancy and childbirth. Freud believed that women never forgive their mothers for leaving them "castrated" and without a penis. Certainly Freud's little sister was left with something to think about after her mother left her without a piano. Freud, the family tyrant, disliked music, and his wish-fulfilments were law.

Pentheraphobia Fear of mother-in-law. (Novercaphobia).

Perceived behavioural control In ajzen's theory of planned behaviour, a component which refers to the person's perceptions of how easy or difficult it would be for them to perform the behaviour, and which takes account of past behaviour and situational and resource factors.

Perceived control The belief that one has the ability to make a difference in the course or the consequences of some event or experience; often helpful in dealing with stressors.

Perceived locus of control A person's belief about the source of outcomes that befall her. That perceived source (locus) may be internal (the result of something she did) or external (the result of forces out of her control).

Percentile rank Percentage of scores falling at or below a specific score. A percentile rank of 95 means that 95% of all of the scores fall at or below this point. In other words, the score is as good as or better than 95% of the scores.

Percentile score A score that indicates the percentage of people who achieved the same as or less than a particular score.

Perception The processes that organize information in the sensory image and interpret it as having been produced by properties of objects or events in the external, three-dimensional world.

Perceptual adaptation The gradual adjustment to various distortions of the perceptual world.

Perceptual constancy The ability to perceive objects as unchanged despite the change noticed by the senses (e.g., the ability to understand and see buildings as remaining the same height even though they appear larger as we get closer to them).

Perceptual hypothesis The perceiver's assumption about what the stimulus is, tested as the perceptual system analyzes the stimulus for appropriate features.

Perceptual model An internal representation of the world, based upon prior knowledge, that can explain the current pattern of physical stimulation. Periodic waveform a repetitive waveform consisting of a sequence of identical cycles.

Perceptual organization The processes that put sensory information together to give the perception of a coherent scene over the whole visual field.

Perceptual parsing The process of grouping various visual elements of a scene appropriately, deciding which elements go together and which do not.

Perceptual set The readiness to see in a particular way that's based on expectations, experiences, emotions, and assumptions.

Perceptual speed The amount of time a person takes to accurately perceive and discriminate between stimuli.

Peremptory challenge A means by which lawyers can exclude a limited number of prospective jurors without the judge's approval.

Perfect correlation A correlation of either +1.0 or -1.0. A perfect correlation is extremely rare and when it occurs means that predicting one score based on another score is perfect or without error.

Performance appraisal The process of evaluating an employee's work within the organization.

Performance assessment Assessment of a student's ability to perform tasks, not just knowledge.

Performance goals A motivational orientation of students who place primary emphasis on gaining recognition from others and earning good grades.

Perineum The region of the body between the thighs, at the outlet of the pelvis; specifically, the small area between the anus and the vulva in the female or between the anus and the scrotum in the male.

Period of concrete operations The third period in Piaget's theory of cognitive development, during which children come to understand the conservation principle and other concepts, such as categorization.

Period of formal operations The fourth period in Piaget's theory of cognitive development, during which individuals first become capable of more formal kinds of abstract thinking and hypothetical reasoning.

Peripheral nervous systems (PNS) A series of systems, which together with the central nervous system forms the two major divisions of the body's nervous system. They are located outside the bony core of the skull and spine and are in turn divided into two systems: the somatic and the autonomic.

Peripheral route to persuasion The path of persuasion when the issue being discussed does not matter much to the listener or the listener is distracted. Relatively superficial factors (whether the person offering the arguments is attractive, the number of arguments offered) will determine persuasiveness. Convincing arguments are less important.

Periphery In vision, the area toward the outside of the retina that has a high concentration of rods, is sensitive to dim light, and is responsible for colourless sensations.

Perlocution In speech act theory, the effect of the act beyond the intention of the speaker, i.e. The things that may happen in a listener's head that the speaker cannot predict.

Permastore Near-permanent retention of some kinds of items in memory, mostly involving semantic or general knowledge (e.g., multiplication tables, names of family members).

Permissive parents Parents who demonstrate high parental warmth and low parental control when interacting with their children.

Permissive pattern A parental style in which parents try not to assert their authority and impose few restrictions or demands on their children.

Persecutory delusion A belief centered on the idea that one is being oppressed, pursued, or harassed.

Perseverance The tendency to continue to perform a behaviour even when it is not being reinforced.

Perseveration The tendency to repeat the same response inappropriately, typically accompanying the defects in strategy formation often observed with prefrontal lesions.

Person centred therapy The therapeutic technique based on humanistic theory which is non-directive and empathic.

Person perception The process of forming impressions about other people.

Person variables Individual differences in cognition, which, according to Mischel, include competencies, encoding strategies and personal constructs, expectancies, subjective values, and self-regulatory systems and plans.

Person/client centred psychotherapy Person centred psychotherapy was developed by Carl Rogers. He referred to it as counselling rather than psychotherapy. He also believed that the relationship between the client and the therapist is not a patient-doctor relationship in which the patient passively submits to something that is done to him/her by the healer. On the contrary, it should be a person-to-person relationship in which the therapists talks with the client. By using the word "client" instead of "patient," Rogers wanted to indicate that the client is not sick in any organic sense.

Personal attribution Attribution to internal characteristics of an actor, such as ability, personality, mood, or effort.

Personal distress An unpleasant state of arousal in which people are preoccupied with their own emotions of anxiety, fear, or helplessness upon viewing a victim's plight.

Personal norms In the theory of reasoned action, those determinants of behaviour such as feelings of moral obligation.

Personal space The buffer zone we like to maintain around our bodies. Its size depends on our familiarity with whoever is near us.

Personal unconscious An individual's unconscious, unique to him or her.

Personality The set of habitual attitudes and styles of interacting with people and the world around us which the child acquires as part of its development. Persuasive communications in the theory of attitude change, those messages that attempt to change people's minds despite being different from the participants' initial attitudes. Phase change in environmental cognition, the boundaries on a map visualised by an individual, such as where land gives way to water. Phase locking the response of auditory neurones in synchrony with the stimulus and which forms the basis of the time theory of pitch perception. Phenomenology in the study of personality, a method of studying conscious, mental experiences in relation to specific events. Phoneme a basic 'atomic' speech sound. Photopic vision daylight vision, mediated primarily by cones and therefore preserving colour information.

Personality disorder A chronic, inflexible, maladaptive pattern of perceiving, thinking, and behaving that seriously impairs an individual's ability to function in social or other settings.

Personality inventory A self-report questionnaire used for personality assessment that includes a series of items about personal thoughts, feelings, and behaviours.

Personality psychology The branch of psychology that attempts to categorize and understand the causes of individual differences in patterns of behaviour.

Personality regression Going back to a previous, immature level of personality, deterioration.

Personality tests Tests that measure the non-cognitive parts of human personality, such as interests, values, and personality traits.

Personality trait An enduring personal characteristic that reveals itself in a particular pattern of behaviour in a variety of situations.

Personality types Different categories into which personality characteristics can be assigned based on factors such as developmental experiences or physical characteristics.

Personation An offenders ritualistic actions. Seen with body positioning, mutilation, and other symbolic gestures. This behaviour is only of significance to the offender.

Person-by-situation interaction A view of influences on personality which holds that a person's behaviour is the joint outcome of his predispositions and the particular situations he encounters. Thus, some people may on average be equally fearful, but while one is afraid of meeting people but unafraid of large animals, another may be afraid of large animals but be unafraid of meeting people.

Person-centred theory A theory, proposed by Carl Rogers, stating that the self-concept is the most important feature of personality.

Persuasion The deliberate attempt to influence the thoughts, feelings or behaviours of another.

Persuasive communications In the theory of attitude change, those messages that attempt to change people's minds despite being different from the participants' initial attitudes.

Perversion A sexual drive component that fails to come under the dominance of the genital area and its reproductive needs and instead focuses on some non-reproductive object or aim.

Pessimist A person who expects the worst.

Pessimistic Expecting the worst in this worst of all possible worlds.

PET (positron emission tomography) scan A technique for examining brain function by observing the degree of metabolic activity of different regions of the brain.

Petting Sexual activities other than intercourse.

Phagophobia Fear of swallowing or of eating or of being eaten.

Phalacrophobia Fear of becoming bald.

Phallic personality Stemming from the Phallic stage, a child who becomes fixated may develop a personality characterized by selfishness, impulsivity, and a lack of or reduced ability to feel empathy.

Phallic stage The third of Freud's psychosexual stage, during which the primary erogenous zone is the genital area. During this time, children become attached to the opposite-sex parent.

Phallophobia Fear of a penis, esp erect.

Phantom limb phenomenon As experienced by amputees, extreme or chronic pain in a limb that is no longer there.

Pharmacophobia Fear of taking medicine.

Pharmacotherapy The treatment of psychological problems with chemical agents.

Phase change In environmental cognition, the boundaries on a map visualised by an individual, such as where land gives way to water.

Phase locking The response of auditory neurones in synchrony with the stimulus and which forms the basis of the time theory of pitch perception.

Phasmophobia Fear of ghosts.

Phengophobia Fear of daylight or sunshine.

Phenomenology A philosophy that puts experience above conceptualizations about it. For Jung, some implications: all we ever experience comes through the filter of the psyche and is therefore psychological; that being so, we can never directly know of anything beyond the psyche; and psychological experiences are as real as external objects and not reducible to other (deduced) properties. There are really no fixed principles or valid judgments, but only sheer experience, and at this level (but not below it) psychology must abdicate as a science. "Just as the discovery of radioactivity overthrew the old physics and necessitated a revision of many scientific concepts, so all disciplines that are in any way concerned with the realm of the psychic are broadened out and at the same time remoulded by depth psychology.".

Since the middle of the Eighteenth Century, "Phnomenologie," like its English equivalent, has been a name for several disciplines, an expression for various concepts. Lambert, in his Neue Organon (1764), attached the name "Phnomenologie" to the theory of the appearances fundamental to all empirical knowledge. Kant adopted the word to express a similar though more restricted sense in his Metaphysische Anfangsgrnde der Naturwissenschaft (1786). On the other hand, in Hegel Phnomenologie des Geistes (1807) the same word expresses a radically different concept. A precise counterpart of Hegel's title was employed by Hamilton to express yet another meaning. In "The Divisions of Philosophy" (Lectures on Metaphysics, 1858), after stating that "Philosophy properly so called" is "conversant about Mind," he went on to say: "If we consider the mind merely with the view of observing and generalizing the various phaenomena it reveals, . . . we have . . . one department of mental science; and this we may call the Phaenomenology of Mind." Similarly Moritz Lazarus, in his Leben der Seele (1856-57), distinguished Philnomenologie from Psychologie: The former describes the phenomena of mental life; the latter seeks their causal explanation.

Edmund Husserl (1859-1938) was the first to apply the name "Phnomenologie" to a whole philosophy. His usage, moreover, has largely determined the senses commonly attached to it and cognate words in the Twentieth Century. In his Logische Untersuchungen (1900-01), Husserl gave the name to such investigations and theories as make up most of that work and of the only published volume of his Philosophie der Arithmetik (1891). This established what was to remain the primary denotation of the term in all his later writings. On the other hand owing to changes in his concept of his unchanging theme the explicit connotation of the term, as used by him, underwent development and differentiation.

In the first edition of the Logische Untersuchungen, phenomenology was defined (much as it had been by Hamilton and Lazarus) as descriptive analysis of subjective processes Erlebnisse. Thus its theme was unqualifiedly identified with what was commonly taken to be the central theme of psychology; the two disciplines were said to differ only in that psychology sets up causal or genetic laws to explain what phenomenology merely describes. Phenomenology was called "pure" so far as the phenomenologist distinguishes the subjective from the objective and refrains from looking into either the genesis of subjective phenomena or their relations to somatic and environmental circumstances. Husserl's "Prolegomena zur reinen Logik," published as the first part of the Logische Untersuchungen, had elaborated the concept of pure logic, a theoretical science independent of empirical knowledge and having a distinctive theme: the universal categorial forms exemplified in possible truths, possible facts, and their respective components. The fundamental concepts and laws of this science, Husserl maintained, are genuine only if they can be established by observing the matters to which they apply. Accordingly, to test the genuineness of logical theory, "wir wollen auf die 'Sachen selbst' zurckgehen": we will go, from our habitual empty understanding of this alleged science, back to a seeing of the logical forms themselves. But it is then the task of pure phenomenology to test the genuineness and range of this "seeing," to distinguish it from other ways of being conscious of the same or other matters. Thus, although pure phenomenology and pure logic are mutually independent disciplines with separate themes, phenomenological analysis is indispensible to the critical justification of logic. In like manner, Husserl maintained, it is necessary to the criticism of other alleged knowledge; while, in another way, its descriptions are prerequisite to explanatory psychology. However, when Husserl wrote the Logische Untersuchungen, he did not yet conceive phenomenological analysis as a method for dealing with metaphysical problems.The most radical changes in this concept of phenomenology and its relations to other disciplines had taken place before Husserl wrote his Ideen zu einer reinen Phnomenologie und phnomenologischen Philosophie, of which the only published volume, "General Introduction to Pure Phenomenology," appeared in 1913. They resulted from a development having two main aspects.

1.Phenomenological analyses, partly summarized in the Logische Untersuchungen, had led Husserl to the view that material (generic and specific) as well as logically formal universals or essences are themselves observable, though non-individual, objects. Further analyses showed that awareness of an essence as itself presented might be based on either a clear experiencing or a clear phantasying (fictive experiencing) of an example. In either case, the evidence of the essence or eidos involves evidence of some example as ideally possible but not as actual. Consequently, a science like pure logic, whose theme includes nothing but essences and essential possibilities,—in Husserl's later terminology, an eidetic science—involves no assertion of actual existence. Husserl used these views to redefine phenomenology itself. The latter was now conceived explicitly as the eidetic science of the material essences exemplified in subjective processes, qua pure possibilities, and was accordingly said to be pure also in the way pure logic is pure. A large proportion of the emendations in the second edition of the Logische Untersuchungen serve to clarify this freedom of phenomenology from all presuppositions of actual individual existence—particularly, psychic existence.

2. Under the influence of Franz Brentano (1838-1917), Husserl coined the name "Intentionalitt" for what he saw as the fundamental character of subjective processes. The reflectively experiencable part of one's stream of consciousness is, on the one hand, consciousness of subjective processes as immanent in the stream itself and, on the other hand, consciousness of other objects as transcending' the stream. This character of subjective processes as consciousness of, as processes in which something is intended, is a property they have intrinsically, regardless of whether what is intended in them exists. Seeing intentionality as the fundamental attribute of subjective processes, Husserl held that phenomenology must describe them not only with respect to their immanent components but also with respect to their intended objects, as intended; in the language of his Ideen, phenomenological description must be "noematic," as well as "noetic" and "hyletic.".

Every conscious process intends its objects as in a context with others, some intended as presented, others intended as to become presented if intended future consciousness takes a particular course. In other words, consciousness is always an intentional predelineating of processes in which objects will be intended, as the same or different within an all-inclusive objective context: the world. A pure phenomenology should therefore describe not only particular intended objects but also the intended world, as intended—as part of the "noematic-objective" sense belonging to consciousness by virtue of the latter's intrinsic intentionality. To be sure, in such noematic-objective description the phenomenologist must still disregard the actual relations of the described subjective processes to other entities in the world. But, Husserl contended, when one disregards everything except the intrinsic nature of subjective processes, one still can see their intentionality; therefore all the entities and relations from which one has abstracted can—and should—reappear as noematic-intentional objects, within one's isolated field. In particular, the disregarded status of the observed stream of consciousness itself, its status as related to other entities in the world, reappears—as a noematic-objective sense which the observed consciousness intends. Moreover, as purely eidetic, phenomenology finds that the intrinsic character of any actual consciousness, as intending a world and itself as in that world, is an essentially necessary determination of any possible consciousness.

Husserl noted, however, that even when one's analyses are thus pure, both abstractively and eidetically, one naturally takes it for granted that possible consciousness is possible in some (otherwise indefinite) possible world. That is to say, besides finding "the world" as part of the intentional objective sense posited in the consciousness under investigation, the investigator continues to apprehend this consciousness as essentially worldly, even though he successfully disregards even its possible relations to other worldly objects. At this point, what Husserl considered as the philosophically decisive change in his concept of phenomenology ensues.

Before writing the Ideen, he had come to believe that, as the reflective observer of one's subjective processes, one can establish and maintain the attitude of a mere onlooker, who does not participate even in his own natural attitude of believing in a possible world and apprehending his consciousness as essentially possible in that world. If this attitude of self-restraint (epoch) is consistently maintained, one can discriminate a status of one's consciousness more fundamental than its actuality or its possibility in a world and one can see that this essential worldliness of consciousness is a reflexive consequence of its more fundamental character as consciousness of a world. One can then see, furthermore, that every intendable object is essentially, and most fundamentally, a noematicintentional object (a phenomenon) and has its being and nature because consciousness regardless of the latter's secondary status as in the world—is intrinsically an (actual or potential) intending of that object, in a certain manner, as having certain determinations. Such was Husserl's contention.

In the Ideen and in later works, Husserl applied the epithet "transcendental" to consciousness as it is aside from its (valid and necessary) self-apperception as in a world. At the same time, he restricted the term "psychic" to subjectivity (personal subjects, their streams of consciousness, etc.) in its status as worldly, animal, human subjectivity. The contrast between transcendental subjectivity and, worldly being is fundamental to Husserl's mature concept of pure phenomenology and to his concept of a universal phenomenological philosophy. In the Ideen, this pure phenomenology, defined as the eidetic science of transcendental subjectivity, was contrasted with psychology, defined as the empirical science of actual subjectivity in the world. Two antitheses are involved, however: eidetic versus factual, and transcendental versus psychic. Rightly, they yield a four-fold classification, which Husserl subsequently made explicit, in his formale und transzendentale Logik (1929), Nachwort zu meinen Ideen (1930), and Mditations Cartsiennes (1931). In these works, he spoke of psychology as including all knowledge of worldly subjectivity while, within this science, he distinguished an empirical or matter-of-fact pure psychology and an eidetic pure psychology. The former is "pure" only in the way phenomenology, as explicitly conceived in the first editranscendental subjects. As one can describe one's actual psychic subjectivity, so one can describe one's actual transcendental subjectivity and thus produce an empirical transcendental phenomenology. Again, as in the case of the purely psychic, so in the case of the purely transcendental, an eidetic reduction enables one to tion of the Logische Untersuchungen, is pure: actual psychic subjectivity is abstracted as its exclusive theme; objects intended in the investigated psychic processes are taken only as the latter's noematic-intentional objects. Such an abstractive and self-restraining attitude, Husserl believed, is necessary, if one is to isolate the psychic in its purity and yet preserve it in its full intentionality. The instituting and maintaining of such an attitude is called "psychological epoch"; its effect on the objects of psychic consciousness is called "psychological reduction." As empiricism, this pure psychology describes the experienced typical structures of psychic processes and of the typical noematic objects belonging inseparably to the latter by virtue of their intrinsic intentionality. Description of typical personalities and of their habitually intended worlds also lies within its province. Having acquired empirical knowledge of the purely psychic, one may relax one's psychological epoche and inquire into the extrapsychic circumstances under which, e.g., psychic processes of a particular type actually occur in the world. Thus an empirical pure intentional psychology would become part of a concrete empirical science of actual psychophysical organisms.

If the psychologist, having isolated some instance of subjectivity, considers it only as a purely possible example of subjectivity in some possible world, he is effecting a further, socalled eidetic, reduction of the psychic and is in the position to develop an eidetically pure phenomenological psychology or (as Husserl also called it) an eidetic psychological phenomenology. He can discover, not merely empirical types but essential psychic possibilities, impossibilities, and necessities, in any possible world. Moreover, eidetic reduction can be performed, not only on the psychic but also on any other abstractible region of the world, e.g., the physical, the concretely psycho-physical, the cultural. We can develop purely eidetic sciences of every material region (material ontologies), an eidetic science of the formally universal region, "something or other" (formal ontology, the formal logic of possible being), and finally an all-embracing science of the essential (formal and material) compossibilities and non-compossibilities in any possible concrete world. An eidetic psychological phenomenology would thus become coordinated in a universal eidetic science of worldly being.

There is yet a third kind of epoh that allegedly enables one to discriminate subjectivity qua transcendental—by effecting yet another kind of reduction, which Husserl eventually called "transcendental-phenomenological." (In his Ideen he called it simply "phenomenological.") By refraining from participation in one's inveterate (and justifiable) natural attitude of presupposing the world and the status of one's subjectivity in the world, one can see the world (and whatever else one may intend) as fundamentally a noematic-intentional object for transcendental subjectivity—for one's individual self, the subject whose life is one's own transcendental stream of consciousness, and for other produce a purely eidetic science—here an eidetic transcendental phenomenology, the theme of which is the absolutely universal domain of transcendental subjectivity in general, including the latter's noematic-objective sense: the entire world and all its possible variants. This eidetic transcendental phenomenology is what Husserl ordinarily meant when, in the Ideen or subsequent works, he spoke simply of "phenomenology.".

Because the difference between phenomenological pure psychology and transcendental phenomenology depends on a difference in attitude towards "the same" subject-matter, their contents are widely analogous. Husserl maintained, however, that genuine philosophy is possible only as transcendental phenomenology, because it alone is knowledge of that non-worldly nucleus of subjectivity in which everything intendable as immanent or as transcendent is constituted (produced, generated) as an essentially intentional object. As envisaged in the Ideen and later works, phenomenological analysis is chiefly "transcendental-constitutional" analysis of the subjective structures in which the concrete individual world is built up as an intersub-jectively valid transcendent sense for transcendental subjectivity. In the course of such analysis, every legitimate philosophical problem must find its definitive solution. From the transcendentalphenomenological standpoint, however, one traditional problem, namely the relation between what are essentially objects of consciousness and "things-in-themselves" that are not essentially objects of consciousness, is seen to be spurious. On the one hand, it is evidently false that all directly presented objects of consciousness are immanent in the mind; on the other hand, the concept of an entity that is not an intentionally constituted object of transcendental consciousness is evidently self-contradictory. This is the central thesis of what Husserl called his "transcendental-phenomenological idealism.".

The diversity of concepts that Husserl himself expressed by the word "phenomenology" has been a source of diverse usages among thinkers who came under his influence and are often referred to as "the phenomenological school." Husserl himself always meant by "phenome-nology" a science of the subjective and its intended objects qua intentional; this core of sense pervades the development of his own concept of phenomenology as eidetic, transcendental, constitutive. Some thinkers, appropriating only the psychological version of this central concept, have developed a descriptive intentional psychology—sometimes empirical, sometimes eidetic-under the title "phenomenology." On the other hand, Husserl's broader concept of eidetic science based on seeing essences and essentially necessary relations especially his concept of material ontology has been not only adopted but made central by others, who define phenomenology accordingly. Not uncommonly, these groups reject Husserl's method of transcendental-phenomenological reduction and profess a realistic metaphysics. Finally, there are those who, emphasizing Husserl's cardinal principle that evidence-seeing something that is itself presented—is the only ultimate source of knowledge, conceive their phenomenology more broadly and etymolo-gically, as explication of that which shows itself, whatever may be the latter's nature and ontological status.

Phenomenon Any state or process known through the senses rather than by intuition or reasoning.

Phenothiazines A kind of classical antipsychotic medication, such as Thorazine, that seems to be effective in alleviating the major positive signs and symptoms of schizophrenia.

Phenotype The observable characteristics of an organism, resulting from the interaction between the organism's genotype and its environment.

Phenylalanine An amino acid that cannot be transformed due to an enzyme deficiency in those with phenylketonuria (PKU). In an infant with PKU, phenylalanine is converted into a toxic agent that accumulates in an infant's bloodstream and damages the developing nervous system.

Phenylketonuria (PKU) A condition in which one lacks the gene that enables one to metabolize phenylalanine. If detected early enough, this condition can be treated by means of a special diet. If not detected early, this disorder can cause a severe form of retardation.

Pheromones Special chemicals secreted by many animals that trigger particular reactions in members of the same species. Humans seem to have a pheromone that regulates the timing of menstruation in females.

Phi Correlation A correlational technique used when both variables are binary (Example true/false, yes/no, or on/off).

Phi phenomenon An illusion of movement that arises when a series of images is presented very quickly one after another; also called *stroboscopic movement*.

Philemaphobia or Philematophobia Fear of kissing.

Philophobia Fear of falling in love or being in love.

Philosophobia Fear of philosophy.

Philosophy of mind Philosophy of mind is the philosophical study of the nature of the mind, mental events, mental functions, and consciousness.

Phobia An intense fear of a specific object or situation. Most of us consider ourselves to have phobias, but to be diagnosable, the fear must significantly restrict our way of life.

Phobic disorder An unrealistic, excessive fear of a specific class of stimuli that interferes with normal activities. The object of the anxiety is readily identifiable: It may be a snake, an insect, the out-of-doors, or closed spaces.

Phobophobia Fear of phobias.

Phoneme The smallest significant unit of sound in a language. In English, it corresponds roughly to a letter of the alphabet (e.g., *apt, tap,* and *pat* are all made up of the same phonemes).

Phonemic encoding A way of encoding verbal information that emphasizes how words sound.

Phonetic reading Reading by decoding the phonetic significance of letter strings; "sound reading.".

Phonological dyslexia A reading disorder in which people can read familiar words but have difficulty reading unfamiliar words or pronounceable nonwords because they cannot sound out words.

Phonological short-term memory Short-term memory for verbal information.

Phonology The rules in a language that govern the sequence in which phonemes can be arranged.

Phonophobia Fear of noises or voices or one's own voice; of telephones.

Photoaugliaphobia Fear of glaring lights.

Photophobia Fear of light.

Photopic vision Daylight vision, mediated primarily by cones and therefore preserving colour information.

Photopigment A complex molecule found in photoreceptors; when struck by light, it bleaches and stimulates the membrane of the photoreceptor in which it resides.

Photoreceptor A receptive cell for vision in the retina; a rod or a cone.

Phrasal category Lexical categories which can themselves be clumped together, such as determiner-adjective-noun, and are named after the word that is considered to be the most important lexical category - called the head of the phrase - in this case, a noun phrase.

Phrase A sequence of words within a sentence that functions as a unit.

Phrase structure description A tree diagram that shows the hierarchical structure of a sentence. The descending branches of the tree correspond to smaller and smaller units of sentence structure.

Phrase structure grammar (PSG) A set of rules which captures how phrasal categories can be combined to produce grammatically correct sentences, the structures of which are usually illustrated graphically as parse trees.

Phrase structure The organization of sentences into phrases. Surface structure is the phrase organization of sentences as they are spoken or written. Underlying structure is the phrase organization that describes the meaning of parts of the sentence, such as doer, action, and done-to.

Phrenology An early nineteenth-century fad that involved palpating bumps and indentations on the head in order to judge the examinee's intellectual and personality traits. A forerunner of modern theories of cerebral localization, phrenology nonetheless had no validity.

Phronemophobia Fear of thinking.

Phthiriophobia Fear of lice. (Pediculophobia).

Phthisiophobia Fear of tuberculosis.

Physical attractiveness stereotype The belief that physically attractive individuals possess socially desirable personality traits and lead happier lives than less attractive persons.

Physical dependence Addiction based on a need to avoid withdrawal symptoms.

Physical development The bodily changes, maturation, and growth that occur in an organism starting with conception and continuing across the life span.

Physical disability Any physical defect, change, difficulty, or condition that has the potential to disrupt daily living.

Physical intimacy Mutual affection and sexual activity.

Physical, astral,causal Different planes of consciousness. The astral plane and causal plane exist in between the physical plane and the absolute state of pure consciousness.

Physiological dependence The process by which the body becomes adjusted to and dependent on a drug.

Physiological psychology The branch of psychology that studies the physiological basis of behaviour.

Pictorial cues The monocular depth cues (such as, interposition, linear perspective, and relative size) that the eye exploits as an optical consequence of the projection of a three-dimensional world on a flat surface.

Pidgin An invented language with an apparently reduced range of style, grammar and lexicon which is formed when two or more cultures, particularly in a colonial setting, come into contact and have to communicate with a mixture of their native languages.

Piloerection Erection of the hairs on the surface of the skin, used by furry animals to conserve heat and by some animals (e.g., cats) as a threat display. In humans, piloerection is manifest as "goosebumps," although, since humans lack full body hair, the response is largely vestigial.

Pineal gland A gland that secretes melatonin.

Pinna The external, visible part of the outer ear.

Pitch The psychological dimension of sound that corresponds to frequency; as frequency increases, pitch appears to rise.

Pituitary gland An endocrine gland that is actually a functional extension of the hypothalamus. The pituitary gland is often called the master gland because many of its secretions trigger hormone secretions in other glands.

Place theory A theory of pitch proposed by Hermann von Helmholtz which states that different regions of the basilar membrane in the cochlea respond to different sound frequencies. The nervous system interprets the excitation from different basilar regions as different pitches.

Placebo An inert substance that cannot be distinguished from a real medication by the patient or subject; used as the control substance in a single-blind or double-blind experiment.

Placebo control An experimental condition in which treatment is not administered; it is used in cases where a placebo effect might occur.

Placebo effect Improvement due to expectation rather than the treatment itself; can occur when participants receive a treatment that they consider likely to be beneficial.

Placebo therapy A therapy independent of any specific clinical procedures that results in client improvement.

Placelessness In environmental psychology, the lack of purposeful encounter through transactions with particular environments.

Placenta A vascular organ that is connected to the embryo by the umbilical cord and that is discharged shortly after birth; the structure serves to provide nourishment for and eliminate wastes from the foetus.

Placental lactogen A hormone produced by the placenta that prepares the mammary glands to secrete milk.

Placophobia Fear of tombstones.

Plasticity The ability of the brain, especially in our younger years to compensate for damage.

Platykurtic A curve or distribution of scores that has a lot of variance.

Pleasure center According to some theorists, a special region of the brain that is triggered whenever any motive is satisfied.

Pleasure principle (Long version: the Pleasure Pain Principle) Our most fundamental striving is toward pleasure and away from pain. Pleasure is what we feel when some kind of tension is relieved. Two types: forepleasure (infantile stimulation) and endpleasure (mature satisfaction).

Pluralistic ignorance The state in which people mistakenly believe that their own thoughts and feelings are different from those of others, even when everyone's behaviour is the same.

Plutophobia Fear of wealth.

Pluviophobia Fear of rain or of being rained on.

Pneumatiphobia Fear of spirits.

Pnigophobia or pnigerophobia Fear of choking of being smothered.

Pocrescophobia Fear of gaining weight. (Obesophobia).

Pogonophobia Fear of beards.

Poinephobia Fear of punishment.

Point biserial correlation A correlational technique used when one variable is numeric and the other is binary (Example age and sex or income and true/false).

Point estimation Estimating the population statistic based on a single sample statistic.

Polarization .1 in the nerve cell, the electrical potential difference across the cell wall which enables it to transmit a signal along itself or to another cell. .2 in social psychology, the phenomenon whereby the effect of group discussion of an issue serves to leave individuals with an expressed view which is slightly more extreme than the one they expressed beforehand.

Poliosophobia Fear of contracting poliomyelitis.

Politicophobia Fear or abnormal dislike of politicians.

Polyandry A type of polygamous mating system in which one female monopolizes the reproductive efforts of several males.

Polygamy Any mating system, including polyandry and polygymy, in which one member of a sex monopolizes the reproductive efforts of several members of the other sex.

Polygenic inheritance Inheritance of an attribute whose expression is controlled not by one but by many gene pairs.

Polygenic traits Traits influenced by several genes.

Polygraph A device for measuring heart rate, respiration, and galvanic skin response. Sometimes called a "lie detector," the polygraph can only detect signs of physiological stress; whether one is lying is an interpretation made by the polygraph examiner.

Polygynandry The mating of several females with several males.

Polygyny A type of polygamous mating system in which one male monopolizes the reproductive efforts of several females.

Polymorphous perversity The young child's tendency to get sexual gratification from damn near anything, regardless of its gender or nature. As a result of **psychosexual development**, this tendency is usually replaced by "normal" heterosexuality.

Polyphobia Fear of many things.

Ponophobia Fear of overworking or of pain.

Pons An enlarged section towards the top of the brain containing groups of neurones that are involved in processes like sleep and arousal, motor control, and pain sensation and control.

Poorna chakra Diagram designed by George Mathew to show the apparent formation and dissolution of mind.

Population A body of persons having qualities or characteristics in common.

Population mean The true mean of the entire population often estimated using the sample mean. Abbreviated with the lowercase Greek letter mu. (m).

Population standard deviation The true standard deviation of the population often estimated by using the sample standard deviation. Often abbreviated with the lowercase Greek letter sigma. (s).

Pornography The combination of sexual material with abuse or degradation in a manner that appears to endorse, condone, or encourage such behaviour.

Porphyrophobia Fear of the colour purple.

Portfolio assessment Assessment of a collection of the students work in an area showing growth, self-reflection, and achievement.

Positive correlation A correlation where as one variable increases, the other also increases, or as one decreases so does the other. Both variables move in the same direction.

Positive manifold All-positive correlation matrix, which results in (but is not identical to) a general factor.

Positive punishment Procedure of adding an event or thing to a situation contingent on the occurrence of a behaviour. Probability of occurrence (POO) of the related behaviour decreases.

Positive reinforcement Procedure of adding an event or thing to a situation contingent on the occurrence of a behaviour. Probability of occurrence (POO) of the related behaviour remains unchanged or increases.

Positive reinforcer Consequence given to strengthen behaviour.

Positive skew A frequency curve in which the mean is higher than the median, and the median is higher than the mode. The result is a "tail" towards the more positive side of the graph.

Positive symptoms Symptoms indicated by the presence of altered behaviours.

Positive symptoms of schizophrenia Symptoms that involve behaviour or thinking that is either less pronounced or nonexistent in normal individuals, such as hallucinations, delusions, or bizarre behaviour.

Positively skewed distribution A data distribution with a few very high scores.

Positivity bias The tendency for people to evaluate individual human beings more positively than groups or impersonal objects.

Positron emission tomography (PET) A method of obtaining detailed pictures of activity in the living brain. Involves injecting a radioactive substance that is taken up by the active neurons, and allows one to view the blood flow in localized areas of the brain as certain tasks are preformed.

Possible selves The ideal selves that a person would like to become, the selves a person could become, and the selves a person is afraid of becoming; components of the cognitive sense of self.

Post hypnotic amnesia Forgetting selected events due to suggestion while under hypnosis, including the suggestion itself.

Post traumatic stress disorder A common anxiety disorder that develops after exposure to a terrifying event or ordeal in which grave physical harm occurred or was threatened.

Postconventional level of morality Stages 5 and 6 in Kohlberg's model of moral development, in which individuals make moral judgements in relation to abstract principles.

Postconventional morality Moral reasoning and behaviour characterized by accepting the relative and changeable nature of rules and laws, and conscience-directed concern with human rights.

Postformal thinking The objective use of practical common sense to deal with unclear problems.

Posthypnotic suggestibility The tendency of a person to perform a behaviour suggested by the hypnotist some time after the person has left the hypnotic state.

Postmature baby An infant who is born 2 or more weeks after its due date.

Postpartum stage The period following childbirth.

Postsynaptic membrane The membrane of the receiving cell across the synaptic gap that contains specialized receptor sites.

Postsynaptic neuron A neuron with which the terminal buttons of another neuron form synapses and that is excited or inhibited by that neuron.

Postsynaptic potential The voltage change that occurs at a receptor site of a postsynaptic neuron when a neurotransmitter molecule links up with a receptor molecule.

Posttraumatic stress disorder (PTSD) An anxiety disorder characterized by the persistent reexperience of traumatic events through distressing recollections, dreams, hallucinations, or dissociative flashbacks; develops in response to rapes, life-threatening events, severe injuries, and natural disasters.

Postural echo A channel of communication, often evident in interview situations, where the interviewee will unconsciously shift position as the interviewer does.

Potamophobia Fear of rivers or running water.

Potentiation In motivation, the tendency to make some behaviours, perceptions, and feelings more probable than others.

Potophobia Fear of alcohol.

Power The strength or the data to find a difference when there truly is a difference. Power is abbreviated with the capital Greek letter beta (?).

Power law of practice The theory which states that performance at a particular skill will improve as a simple 'power' function of the amount of practice one has had at the skill.

PQ4R method A study stategy that has students preview, question, read, reflect, recite, and review material.

Practical intelligence The idea of a general cognitive, physical, and spatial awareness that is innate in individuals, and which allows them to navigate spaces and social situations.

Pragmatic reasoning schema An explanation of the effects of content in the selection task which suggests that people use context-dependent rules of inference, such as a permission schema, to solve problems of this kind.

Preadolescence The period of childhood between ages 10 and 11.

Preattentive processing Processing of sensory information that precedes attention to specific objects.

Precision The ability of a measurement to be consistently produced.

Preconscious memories Memories that are not currently conscious but that can easily be called into consciousness when necessary.

Preconscious That which upon reflection could become conscious voluntarily. For instance, remembering your wife's name when you weren't thinking about her. The ego's inner layer. That which is unconscious is repressed and therefore held out of awareness.

Preconventional level of morality Stages 1 and 2 in Kohlberg's model of moral development, in which individuals make moral judgments in their own interests.

Preconventional morality Moral reasoning and behaviour based on rules and fear of punishment and nonempathetic self-interest.

Precursor A substance required for the chemical manufacture of a neurotransmitter.

Predatory aggression Our motivated attack behaviours. This aggression is directed to natural prey and is deeply routed in our ancestors hunting behaviour. Today it can be seen in the behaviour of normal individuals as hunting.

Predicate thinking The id's tendency to treat dissimilar objects as identical.

Predictive validity A measure of whether a test assesses what is intended that is based on the correlation between the test score and some external criterion (e.g., a correlation between a scholastic aptitude test score and college grades).

Prefrontal area The frontmost portion of the frontal lobes, which is involved in working memory, strategy formation, and response inhibition.

Prefrontal cortex The anterior part of the frontal lobe; contains the motor association cortex.

Prefrontal lobotomy A neurosurgical treatment that surgically cuts the connections between the prefrontal areas of the frontal lobes and the rest of the brain. Once used widely (and mostly unsuccessfully) for many mental disorders but now performed very rarely.

Pregnancy The condition, quality, or period of having (an) offspring developing in the uterus.

Prejudice A learned attitude toward a target object, involving negative affect (dislike or fear), negative beliefs (stereotypes) that justify the attitude, and a behavioural intention to avoid, control, dominate, or eliminate the target object.

Premack Principle Using favoured activities to reinforce participation in less desired activities.

Premature (preterm) birth A birth that occurs before a gestation of 37 weeks.

Premise An assumption or stipulation that precedes deductive reasoning.

Prenatal period The nine months between conception and birth. This period is divided into three developmental stages: the zygote, the embryo, and the foetal stages.

Preoperational period In Piaget's theory, the period from about ages two to six during which children come to represent actions and objects internally but cannot systematically manipulate these representations or relate them to each other. The child is therefore unable to conserve quantity across perceptual transformations and also is unable to take points of view other than his own.

Preoptic area A region at the base of the brain just in front of the hypothalamus; contains neurons that appear to control the occurrence of slow-wave sleep.

Preparedness A built-in predisposition to form certain associations more readily than others.

Preparedness theory of phobias The theory that phobias grow out of a built-in predisposition (preparedness) to learn to fear certain stimuli (e.g., snakes and spiders) that may have posed serious dangers to our primate ancestors.

Presbycusis Difficulty hearing high-pitched sounds.

Presbyopia A form of farsightedness occurring after middle age, caused by a diminished elasticity of the crystalline lens.

Prescriptionism The view that psychotherapeutic treatments for mental disorders may ultimately be like prescriptions for medications: tailored to both the disorder and the individual patient.

Prescriptive rules Rules prescribed by "authorities" about how people ought to speak and write that often fail to conform to the facts about how people actually talk and understand. This is in contrast to the structural principles of a language, which describe (rather than prescribe) the principles native speakers of a language actually use when arranging their words into sentences.

Presentation punishment Decreasing the chances that a behaviour will occur again by presenting an aversive stimulus following the behaviour.

Pressure A sense of being compelled to behave in a particular way because of expectations set by oneself or others.

Presynaptic facilitation A process that underlies many kinds of learning, documented in studies of *Aplysia*. It occurs when learning results in the increased readiness of presynaptic neurons to fire.

Presynaptic neuron A neuron whose terminal buttons form synapses with and excite or inhibit another neuron.

Pretend play One of the stages of play amongst children, described by jean piaget and sara smilansky, which children engage in from 15 months onward and which develop from simple acts of 'feeding' dolls into elaborate role-play episodes.

Pretest posttest method A method of determining the amount of change that occurred in a set of data by measuring the data prior to treatment and then after treatment and comparing the two measurement outcomes.

Preventive psychology Any attempt to forestall the development of psychological problems by altering the sociocultural variables predictive of psychological distress.

Primacy effect 1. In free recall, the tendency to recall the first items on a list more readily than those in the middle.
2. In forming an impression of another person, the tendency to give greater weight to attributes noted at the outset than to those noted later.

Primal fantasies Unconscious fantasies stemming from having perceived one's parents having sex.

Primal horde The original primitive grouping of humanity headed by a father-authority who forbad incestuous actings-out among younger men, who, to gain their sexual freedom, eventually overpowered and eating him and thereby discovered the powers and benefits of community. In a way, similar to what befell the Wednesday Psychological Society.

Primary appraisal One of two appraisal processes applied in the study of physiological stress, when an individual appraises the threat inherent in a situation.

Primary auditory cortex The region of the cerebral cortex that receives information directly from the auditory system; located in the temporal lobes.

Primary colour One of a small set of coloured lights which, when added together in different proportions, can produce a wide range of perceivable colours.

Primary defence Most important and commonly used type of psychological device used to protect self-acceptance.

Primary messenger The neurochemicals responsible for neuron-to-neuron communication in chemical synapses, i.e., neurotransmitters. Primary messengers are contrasted with second messengers, those neurochemicals responsible for communication within neurons.

Primary motor cortex The region of the cerebral cortex that directly controls the movements of the body; located in the back part of the frontal lobes.

Primary motor projection area A strip of cortex located at the back of the frontal lobe just ahead of the primary sensory projection area in the parietal lobe. This region is the primary projection area for muscular movements.

Primary process Produces a memory image of an object needed for gratification in order to reduce the frustration of not having been gratified yet. This develops as the id encounters frustrations of its desires, and it works because for the id, an image is the same as an object ("an identity of perception"). Primary process is preverbal and dreamlike, not rational as with the ego. It's called "primary" because it comes first in human development: for Freud if not for later theorists, the baby is, so to speak, all id.

Primary process thinking Thinking that is irrational, illogical, and motivated by a desire of immediate gratification of impulses.

Primary projection areas Regions of the cortex that serve as receiving stations for sensory information or as dispatching stations for motor commands.

Primary punisher A biologically significant aversive stimuli, such as pain.

Primary reinforcer A biologically significant appetitive stimulus, such as food or water.

Primary sex characteristics Any of the physical characteristics differentiating male and female individuals; directly responsible for reproduction.

Primary somatosensory cortex The region of the cerebral cortex that receives information directly from the somatosensory system (touch, pressure, vibration, pain, and temperature); located in the front part of the parietal lobes.

Primary somatosensory projection area A strip of cortex located at the front of the parietal lobe just behind the primary motor area in the frontal lobe. This region is the primary projection area for bodily sensations, including touch, pain, and temperature.

Primary visual cortex In the occipital lobe of the cerebrum, the brain part involved in handling visual information.

Primates A group of mammals, characterized by having flexible hands and feet and highly developed brains, and which includes, humans, apes, lemurs, tarsiers and monkeys.

Priming Activating particular associations in memory.

Priming effect Phenomenon wherein giving a participant advance knowledge about or exposure to a stimulus can increase the ease of its subsequent recall or recognition.

Primitive In biological terms, a term relating to a trait which has manifested itself in the early stages of the evolution of a species.

Primitive features Attributes of an object (such as its location, contour, colour, and shape) that are first detected separately and then coordinated to enable identification of the object.

Principle Explanation of the relationship between factors such as the effects of alternative grading systems on student motivation.

Principle of closure The Gestalt psychology principle that states that people tend to interpret familiar incomplete forms as complete by filling in gaps.

Principle of compositionality A theoretical principle in semantics which indicates that one can combine the basic meanings of all of the words and phrases involved in an utterance to determine the meaning of the whole utterance.

Principle of continuity The Gestalt psychology principle that states that people tend to perceive interrupted lines and patterns as continuous by filling in gaps.

Principle of parsimony The principle of applying the simplest possible explanation to any set of observations; also called *Occam's razor.*

Principle of proximity The Gestalt psychology principle that states that people tend to perceive objects as a group when they are close together.

Principle of similarity The Gestalt psychology principle that states that people tend to group similar objects together.

Principle of simplicity The Gestalt psychology principle that states that people tend to perceive forms as simple, symmetrical figures rather than as irregular ones.

Prison study A famous study done by Philip Zimbardo that showed the influence of roles.

Prisoner's dilemma A particular arrangement of payoffs in a two-person situation in which each individual has to choose between two alternatives without knowing the other's choice. The payoff structure is arranged such that the optimal strategy for each person depends upon whether she can trust the other or not. If trust is possible, the payoffs for each will be considerably higher than if there is no trust.

Private conformity The change of beliefs that occurs when a person privately accepts the position taken by others.

Private self-awareness A psychological state in which one is aware of one's hidden private self-aspects.

Private self-consciousness The tendency to be aware of the covert, private aspects of the self.

Private speech Children's self-talk, which guides their thinking and action. Eventually these verbalizations are internalized as silent inner speech.

Proactive coping Up-front efforts to ward off or modify the onset of a stressful event.

Proactive facilitation Increased ability to learn new information due to previously acquired information.

Proactive inhibition Decreased ability to learn new information because of interference of present knowledge.

Proactive interference Interference in memory due to prior learning.

Proactive interference The forgetting of new information because of previously learned information.

Probability The number of times an event is likely to occur out of the total number of possibilities.

Probability of error The likelihood that error caused the results of data analysis. If the probability of error is greater than the predetermined acceptable level of error then the results are said to be 'not significant.'.

Probability of response The likelihood of the occurrence of a response; it is a common measure of response strength in classical and operant conditioning.

Probability sample Also called representative samples, a probability sample consists of characteristics that are close to the population that they represent.

Problem focused coping Cognitive and behavioural efforts to alter a stressful situation.

Problem reduction In problem solving, a technique using the method of 'divide and conquer' applied to reduce one large problem to several smaller ones.

Problem solving Thinking that is directed toward solving specific problems and that moves from an initial state to a goal state by means of a set of mental operations.

Problem solving assessment Involves organizing, selecting, and applying complex procedures that have at least several important steps or components.

Problem space The elements that make up a problem: the initial state, the incomplete information or unsatisfactory conditions the person starts with; the goal state, the set of information or state the person wishes to achieve; and the set of operations, the steps the person takes to move from the initial state to the goal state.

Procedural memory Memory for how things get done; the way perceptual, cognitive, and motor skills are acquired, retained, and used.

Process loss The reduction in group performance due to obstacles created by group processes, such as problems of coordination and motivation.

Process model of persuasion In the study of attitude change, an account which suggests that the effect of a persuasive communication is determined by a five-stage process of attention, comprehension, yielding, retention and behaviour.

Process schizophrenia According to Bleuler, a form of schizophrenia characterized by a gradual onset and a poor prognosis.

Processing stream A neural pathway thought to be concerned primarily with the analysis of one stimulus attribute, such as form stream, motion stream.

Process-product studies Research approach in which the teaching practices of effective teachers are recorded through classroom observation.

Proctophobia Fear of rectums.

Production systems In problem solving applications, a system for solving problems that is built from a set of 'if...then' rules to be applied to a particular situation.

Productive language An ability to use the spoken or written word.

Progesterone A hormone secreted by the corpus luteum, active in preparing the uterus for the reception and development of the fertilized ovum and the mammary glands for milk secretion.

Prognosis A prediction about the probable course and outcome of a disorder.

Programmed instruction Structured lessons that students can work on individually, at their own pace.

Progressive relaxation technique A relaxation technique involving three steps: 1. recognizing the body's signals that indicate the presence of stress;.

2. Using those signals as a cue to begin relaxing; and.

3. Relaxing groups of muscles, beginning with those in the head and neck and then those in the arms and legs.

Projection In psychoanalytic theory, a mechanism of defence in which various forbidden thoughts and impulses are attributed to another person rather than the self, thus warding off some anxiety (e.g., "I hate you" becomes "You hate me").

Projective hypothesis The idea that people interpret ambiguous stimuli in ways that reveal their concerns, needs, conflicts, desires, and feelings.

Projective personality tests Unstructured personality measures in which a person is shown a series of ambiguous stimuli, such as pictures, inkblots, or incomplete drawings. The person is asked to describe what he or she "sees" in each stimulus or to create stories that reflect the theme of the drawing or picture.

Projective techniques Sometimes called *unstructured personality tests.* Methods of assessing personality that use relatively ambiguous stimuli in order to elicit responses that are unguarded and authentic. The most common projective techniques are the TAT and the Rorschach inkblot test.

Projective test A method of personality assessment in which an individual is presented with a standardized set of ambiguous, abstract stimuli and asked to interpret their meanings; the individual's responses are assumed to reveal inner feelings, motives, and conflicts.

Promiscuity Characterized by a lack of discrimination; specifically, engaging in sexual intercourse indiscriminately or with many persons.

Propagation The spread of the action potential down an axon, caused by successive destabilizations of the neuronal membrane.

Propositional storage and retrieval In environmental cognition, the process of storing spatial information as a string of instructions for movement along a route.

Prosocial behaviour The capacity to help, cooperate, and share with others.

Prosody The use of changes in intonation and emphasis to convey meaning in speech besides that specified by the particular words; an important means of communication of emotion.

Prosopagnosia A form of visual agnosia characterized by difficulty in the recognition of people's faces; caused by damage to the visual association cortex.

Prosopagnosia The inability to recognize faces, usually produced by lesions in the parietal lobes.

Prosophobia Fear of progress.

Protanopia A form of hereditary anomalous colour vision; caused by defective "red" cones in the retina.

Protection The preservation of the participant from any risk of physical or mental harm during the course of a psychological experiment, as well as the ensuring of their privacy during observation.

Protestant ethic A phrase that describes and relates to early American culture, emphasizing individual achievement, personal responsibility, self-sufficiency, and control over the environment.

Protoconversations The earliest form of communication between carer and infant based on an exchange of looks, smiles and sounds which engage the baby's attention.

Prototype A hypothetical idealized pattern that resides in the nervous system and is used to perceive objects or shapes by a process of comparison; recognition can occur even when an exact match is not found.

Prototype theory of meaning The theory that concepts are formed around average exemplars rather than lists of single attributes.

Protoword A unique string of phonemes that an infant invents and uses as a word.

Provocation Others actions that trigger aggression in the recipient because they are seen as stemming from malicious intent.

Proxemics In the study of communication, the use of space between individuals in various social or cultural situations, and which conveys varying degrees of intimacy and friendliness.

Proximal stimulus A near stimulus, that acts directly on an aspect of the nervous system. The effect of the light waves on the retina of the eye is an example of proximal stimulus (see Distal Stimulus).

Proximate cause The immediate cause in a chain of causation. The proximate cause of a person's death might be a stroke, but *why* he suffered a stroke requires further investigation.

Proximity 1. In perception, the closeness of two figures. The closer together they are, the more they will tend to be grouped together perceptually;.

2. The nearness of people, which is one of the most important determinants of attraction and liking.

Psellismophobia Fear of stuttering.

Pseudodiagnosticity In heuristics, the idea that people tend to retain an initial (and often favoured) hypothesis or try to confirm rather than disconfirm hypotheses when making decisions about things.

Pseudohermaphroditism The most common kind of intersexuality, in which individuals have ambiguous genitalia.

Psi capacities Paranormal capacities which transcend and motor limitations of man, namely, ESP (Extra Sensory Perception) and PK(Psycho Kinesis).

Psychedelic drugs Drugs which alter or expand consciousness changing perceptual quality, sense of time, etc.

Psychiatric disability A mental illness or psychological disturbance, causing a person to struggle with mild to incapacitating emotional problems and limitations that are often caused by either anxiety or affective disorders.

Psychiatrist An individual who has obtained an M.D. degree and also has completed postdoctoral specialty training in mental and emotional disorders; a psychiatrist may prescribe medications for the treatment of psychological disorders.

Psychiatry Psychiatry is a branch of medicine that studies and treats mental and emotional disorders, some of which are listed on the mental illness page. The term alienist is an old term for a psychiatrist.

Psychic determinism The assumption that mental and behavioural reactions are determined by previous experiences.

Psychoactive drugs Chemicals that affect mental processes and behaviour by temporarily changing conscious awareness of reality.

Psychoactive substance use disorders Disorders are characterized by addiction to drugs or by abuse of drugs.

Psychoanalysis 1. A theory of both normal and abnormal human personality development, formulated by Freud, whose key assertions include unconscious conflict and early psychosexual development.

2. A method of therapy that draws heavily on this theory of personality. Its main aim is to have the patient gain insight into her own, presently unconscious, thoughts and feelings. Therapeutic tools employed toward this end include free association, interpretation, and the appropriate use of the transference relationship between patient and analyst.

Psychoanalyst An individual who has earned either a Ph.D. or an M.D. degree and has completed postgraduate training in the Freudian approach to understanding and treating mental disorders.

Psychoanalytic model As defined in the text, a subcategory of the pathology model which holds 1. That the underlying pathology is a constellation of unconscious conflicts and defences against anxiety, usually rooted in early childhood, and.

2. That treatment should be by some form of psychotherapy based on psychoanalytic principles.

Psychoanalytic theory Theory developed by Freud consisting of the structural model of personality, topographical model of personality, defence mechanisms, drives, and the psychosexual stages of development. The primary driving force behind the theory is the id, ego and superego and the division of consciousness into the conscious mind, the pre/subconscious, and the unconscious.

Psychoanalytic therapy Freudian technique of exploring unconscious motivations, conflicts, and repressed emotions for a prolonged period of time; an alternative to biomedical treatment.

Psychobiography The use of psychological (especially personality) theory to describe and explain an individual's course through life.

Psychobiology The study of the interrelationships of the biology and psychology in cognitive functioning, including intellectual, memory, and related neurocognitive processes.

Psychodynamic A term used to describe the Freudian notion that the mind is in a state of conflict among instincts, reasons, and conscience.

Psychodynamic approach to personality An approach to personality originally derived from psychoanalytic theory that asserts that personality differences are based on unconscious (dynamic) conflicts within the individual.

Psychodynamic model A theoretical model applied to the diagnosis of mental-health and physical problems psychometric a term applied in intelligence research to cover those aspects concerned with the design and the use of intelligence tests, and related research which has been based on the results of intelligence tests.

Psychodynamic personality theories Theories of personality that share the assumption that personality is shaped by and behaviour is motivated by powerful inner forces.

Psychodynamic perspective A psychological model in which behaviour is explained in terms of past experiences and motivational forces; actions are viewed as stemming from inherited instincts, biological drives, and attempts to resolve conflicts between personal needs and social requirements.

Psychodynamic theories Theories based on the work of Sigmund Freud. These theories emphasize unconscious motives and desires and the importance of childhood experiences in shaping personality.

Psychodynamic therapy A modern adaptation of psychoanalytic therapy which has made sometimes minor and sometimes major changes to Freud's original theories.

Psychogenic amnesia A dissociative disorder characterized by the inability to remember important events or personal information.

Psychogenic disorders Disorders whose origins are psychological rather than organic (e.g., phobias).

Psychogenic fugue Amnesia with no apparent organic cause accompanied by a flight away from home.

Psychogenic symptoms Symptoms believed to result from some psychological cause rather than from actual tissue damage.

Psychokinesis Mind on matter effect. Change of state or motion induced not using the physical motor system of the body.

Psycholinguistics The branch of psychology devoted to the study of verbal behaviour.

Psycholinguists Specialists in the study of language.

Psychological adolescing The process of growing up to full adulthood and realizing the personal potential of oneself; one part of the aging process that develops along with growing older physically.

Psychological androgyny Individuals with both male and female psychological characteristics; a blend of masculinity and femininity that results in a greater behavioural adaptability.

Psychological assessment The use of specified procedures to evaluate the abilities, behaviours, and personal qualities of people.

Psychological autopsy An investigative review and victimology interview procedure used to determine the victims psychological makeup.

Psychological dependence Addiction based on cravings for a drug.

Psychological diagnosis The label given to psychological abnormality by classifying and categorizing the observed behaviour pattern into an approved diagnostic system.

Psychological intensity The magnitude of a stimulus as it is perceived, not in terms of its physical attributes.

Psychological intimacy The sharing of feelings and thoughts.

Psychological reactance The theory that people react against threats to their freedom by asserting themselves and perceiving the threatened freedom as more attractive.

Psychological testing Psychological testing is a field characterized by the use of small samples of behaviour in order to infer larger generalizations about a given individual. The technical term for psychological testing is psychometrics. By samples of behaviour, we mean observations of the individual over a limited amount of time performing tasks which have usually been prescribed beforehand, often with a great deal of research into the responses of members of a norm group. These responses are often compiled into statistical tables that allow the evaluator to compare the behaviour of the individual being tested to the responses of the range of responses given by people in the norm group. When multiple tests are administered, the procedure is referred to as full battery assessment.

Psychology Psychology is an applied discipline and applied profession concerned with the study of and intervention in mental states, processes, and behavioural patterns of humans and, to an extent, of animals (though the study of animal behaviour, ethology, is more often regarded a branch of biology than of psychology). Psychologists also study interactions between individuals and groups of individuals, and between individuals, groups and their environment. Disciplines that are traditionally considered to intersect with psychology are sociology, anthropology, biology, and philosophy, but more recently fields such as neuroscience, political science, media studies and gender studies have also come to be seen as closely related to psychology.

Psychology of consciousness The branch of Psychology which studies behaviour in terms of altered states of consciousness (like waking, sleeping, aesthetic, mystic, etc.).

Psychometric approach A method of understanding intelligence that emphasizes people's performance on standardized aptitude tests.

Psychometric approach to intelligence An attempt to understand the nature of intelligence by studying the pattern of results obtained on intelligence tests.

Psychometric function A graph that plots the percentage of detections of a stimulus (on the vertical axis) for each stimulus intensity (on the horizontal axis).

Psychometric research Studies in the field of psychology that specialize in mental testing and developing standardized methods for collecting data and assessing psychological phenomena.

Psychometric test A performance or iq test often used to assess candidates for jobs.

Psychometrics Psychometrics is the field of study (connected to psychology and statistics) concerned with the measurement of "psychological" aspects of a person such as knowledge, skills, abilities, or personality. The field of Psychometrics is primarily concerned with differences between individuals and employs statistical tools such as normal distribution and factor analysis. Measurement of these unobservable phenomena is difficult and much of the research and accumulated art of this discipline is designed to reliably define and then quantify. Critics, including "hard science" practitioners and social activists, have argued that such definition and quantification is impossibly difficult and that such measurements are very often misused (although users of psychometric techniques can reply that their critics often misuse data by not assessing them with psychometric criteria). Figures who made significant contributions to psychometrics include Karl Pearson, L. L. Thurstone, Georg Rasch and Arthur Jensen. Signi.

Psychoneuroimmunology (PNI) A subfield of psychology that examines the links among psychological factors, the brain and nervous system, and the immune system.

Psychoneuroimmunology The research area that investigates interactions between psychological processes, such as responses to stress, and the functions of the immune system.

Psychopath Former term for a person with antisocial personality disorder.

Psychopathic Pertaining to psychopathy, particularly to antisocial behaviour or antisocial personality disorder.

Psychopathological functioning Disruptions in emotional, behavioural, or thought processes that lead to personal distress or block one's ability to achieve important goals.

Psychopathology In Freudian analysis, a method for gaining insight by analyzing the significance of and underlying motivation for the verbal slips made by patients in conversation.

Psychopathy Criminal behaviour resulting from personality deviations.

Psychopharmacology The branch of psychology that investigates the effects of drugs on behaviour.

Psychophobia Fear of mind.

Psychophysics An approach to understanding perception that relates the characteristics of physical stimuli to attributes of the sensory experience they produce.

Psychophysiological disorders In these disorders (formerly called *psychosomatic*), the primary manifestations involve genuine organic damage, but their onset or severity is heavily influenced by psychological factors (e.g., coronary heart disease).

Psychophysiology The measurement of physiological responses, such as blood pressure and heart rate, to infer changes in internal states, such as emotions.

Psychosexual development Freud's developmental notions, many inspired by Karl Abraham's original work, centered around the idea that libido invests certain bodily zones with energy as the child grows older. The order of stages is: oral (birth to eight months), anal (eight months to two years), phallic (two years to six), latency (six to twelve), genital (puberty to adult). As one masters the stages, one's ability to exert ego control over one's drives grows.

Family conflict can fixate a person on any of these stages and thereby produce a lasting effect on character: the greedy, ever-empty, grasping oral type; the fanatically orderly and controlling anal type; the thrusting, ambitious phallic type; the quiescent latent type (this one isn't seen as so pathological) — and the mature outcome of a successful upbringing OR analysis, the genital type, devoted to love and work, but not necessarily in that order.

Psychosexual development In psychoanalytic theory, the description of the progressive stages in the way the child gains pleasure as he grows into adulthood, defined by the zone of the body through which maximal pleasure is derived (oral, anal, genital) and by the object toward which this pleasurable feeling is directed (mother, father, adult sexual partner).

Psychosis Loss of contact with reality (most often evidenced as delusions or hallucinations), as can occur in severe cases of many kinds of mental disorders such as mania, major depression, or schizophrenia.

Psychosocial Of or pertaining to the psychological development of the individual in relation to his or her social environment.

Psychosocial crises In Erik Erikson's theory, a series of crises through which all persons must pass as they go through their life cycle (e.g., the identity crisis during which adolescents or young adults try to establish the separation between themselves and their parents).

Psychosocial dwarfism A syndrome in which children's normal development is inhibited by traumatic psychological surroundings or events. For instance, physical growth may be stunted when young children experience stress due to family trauma, abandonment, or lack of human touch. Also known as stress dwarfism.

Psychosocial stages Proposed by Erik Erikson, successive developmental stages that focus on an individual's orientation toward the self and others; these stages incorporate both the sexual and social aspects of a person's development and the social conflicts that arise from the interaction between the individual and the social environment.

Psychosocial theory A set of principles that relates social environment to psychological development.

Psychosomatic disorders Physical disorders aggravated by or primarily attributable to prolonged emotional stress or other psychological causes.

Psychosomatics In clinical psychology, the application of psychological research and theories not only to mental health problems but also to physical health problems, by extending the study of clinical problems to primary-care and general hospital settings.

Psychosurgery Neurosurgery performed to alleviate manifestations of mental disorders that cannot be brought under control using psychotherapy, medication, or other standard treatments. Psychosurgery can be helpful in severe cases of, for example, obsessive-compulsive disorder.

Psychotherapy A set of techniques for curing or helping to solve behavioural problems of humans. The common part of these techniques is direct personal contact between therapist and patient.

Psychotic disorders Severe mental disorders in which a person experiences impairments in reality testing manifested through thought, emotional, or perceptual difficulties; no longer used as a diagnostic category after DSM-III.

Psychoticism In Hans Eysenck's personality system, a trait dimension related to aggressiveness, antisocial actions, coldness, impulsivity, and self-centeredness.

Psychotropic medication Prescription medication used primarily to treat mental illness.

Psychrophobia Fear of cold.

Pteromerhanophobia Fear of flying.

Pteronophobia Fear of being tickled by feathers.

Puberty A variable period of a year or so during which the child undergoes relatively rapid development, particularly in its sex organs, and in the case of the female, in its ability to reproduce.

Pubescence The two years before puberty.

Public conformity A superficial change in overt behaviour, without a corresponding change of opinion, produced by real or imagined group pressure.

Public self-awareness A psychological state in which one is aware of one's public self-aspects.

Public self-consciousness A personality characteristic of individuals who focus on themselves as social objects, as seen by others.

Pull-out programs Compensatory education programs in which students are placed in separate classes for remediation.

Punctuated equilibrium In the theory of evolution, a period of brief bursts of rapid change in a species, activated by favourable environmental conditions.

Punisher Event or thing added to or removed from a situation that is contingent on the occurrence of a behaviour. Probability of occurrence (POO) of the related behaviour decreases.

Punishment A consequence that decreases the frequency of a response that is regularly and reliably followed by an aversive stimulus.

Punishment training An instrumental training procedure in which a response is suppressed by having its occurrence followed by an aversive event.

Pupaphobia Fear of puppets.

Pupil An opening that lets light into the back of the eye.

Pure Awareness The absolute state of consciousness, beyond space and time.

Pure tone An alternative word for harmonic or sinusoid.

Pure word deafness The ability to hear, to speak, and (usually) to write, without being able to comprehend the meaning of speech; caused by bilateral temporal lobe damage.

Purified pleasure ego That sector of the ego that rids itself of badness by projecting it onto external objects.

Purposive sampling This is usually done for the purpose of selecting (not randomly) the individuals that will provide the best information for the study.

Pursuit movement The movement that the eyes make to maintain an image upon the fovea.

Puzzle box An apparatus used by Edward Thorndike to demonstrate trial-and-error learning in cats. Animals were required to perform a simple action in order to escape the puzzle box and obtain food.

Pygmalion effect Positive change in a person's performance or perception of a situation, based on expectation and encouragement by others.

Pyramidal system One of the two motor systems that originates in the motor cortex of the brain and sends its tracts directly to the motoneurons that activate the muscles. The pyramidal system is more recent in evolutionary terms and orchestrates the body's more precise movements.

Pyrexiophobia Fear of Fever.

Pyrogens Chemicals released into the bloodstream at sites of bacterial or viral invasion that stimulate special receptors in the anterior part of the hypothalamus and produce fever, which probably represents the body's effort to kill off the invading pathogens by temporarily overheating.

Pyrophobia Fear of fire.

Q

QAIT model A model of effective instruction that focuses on elements that teachers can directly control.

Q-data A category of data used in personality assessment, relating to self-ratings gained from questionnaires.

Qualitative method An increasingly popular method applied by psychologists in reaching insights into human experience and behaviour that cannot be gained from the quantitative method.

Qualitative research Research in which information collected from respondents takes the form of verbal descriptions or direct observations of events.

Quality circles A type of work group, popular in japan and the usa, where small groups of employees from the same work areas meet regularly to identify, analyse and solve quality and other work problems.

Quantitative A quantitative property can be meaningfully measured using numbers; properties which aren't quantitative are called qualitative.

Quantitative method An approach used by psychologists in research studies in order to count quantities and use statistical support in the evaluation of findings.

Quantitative research Research in which information is collected from respondents and converted into numbers.

Quasi experimental A design in which treatments are given but random assignment is not used.

Quasi experimental research Any research study that uses specific experimental methods but does not randomize subjects.

Questionnaire A less personal version of a face-to-face interview, usually intended for a larger collection of subjects but which is nevertheless highly structured in content.

Quickening The stage of pregnancy in which the movement of the foetus can be felt for the first time.

R

R Symbol used for the Pearson-product moment correlation (correlation coefficient).

Race Visible, genetic characteristics of individuals that cause them to be seen as members of the same broad group (e.g. , African, Asian, Caucasian).

Racism 1. An individual's prejudicial attitudes and discriminatory behaviour toward people of a given race, or.

2. Institutional practices (even if not motivated by prejudice) that subordinate people of a given race.

Radical behaviourism An approach usually associated with B. F. Skinner which asserts that the subject matter of psychology is overt behaviour, without reference to inferred, internal processes such as wishes, traits, or expectations.

Radiophobia Fear of radiation, x-rays.

Random assignment The process of assigning participants to the conditions of an experiment such that all persons have the same chance of being in a given condition. (note the distinction between random assignment in experiments and random sampling in surveys. Random assignment helps us infer cause and effect. Random sampling helps us generalize to a population.).

Random dot stereogram A visual stimulus consisting of a pair of random dot patterns whose two members are identical, except that a region in one is displaced relative to the other.

Random sample A sample is a subset of a larger population. A random sample is one chosen by a method involving a random (i.e. Unpredictable) component. A simple random sample is one in which every item/ object therein had an equal opportunity/ chance/probability of being selected from the larger population containing all items/ objects of interest. A probability sample is one in which each item has a known probability of being in the sample.

Randomized Field Experiment Experiment conducted under realistic conditions in which individuals are assigned by chance to receive different practical treatments or programs.

Range The difference between the highest and lowest score in a distribution (often 1 is added to the result when computing statistics to allow for the 0.5 on either end lost due to rounding).

Ranidaphobia Fear of frogs.

Rank ordered array A table consisting of data in order of highest to lowest or lowest to highest where each data is given a numbered rank depicting it's difference from the highest or lowest score.

Rape myth The false belief that deep down, women enjoy forcible sex and find it sexually exciting.

Rapid eye movement (REM) It is the part of the sleep during which people dream. It is named after a characteristic of this part of sleep, which is much different from deep sleep. During a night of sleep, a person usually has about four or five periods of REM sleep, which are quite short at the beginning of the night, but long at the end.

Ratio measurement Much like interval, uses a set scale. The researcher knows the order of the data points both relative to each other and in absolute terms. Both the zero and the average have meaning.

Ratio scale An interval scale in which there is a true zero point, thus allowing statements about proportions (e.g., this sound is twice as loud as the other).

Ratio schedule A reinforcement schedule in which reinforcement is delivered for the first response made after a certain number of responses. In a fixed-ratio schedule, the number of responses required for a reward is always the same. In a variable-ratio schedule, the number of responses required varies irregularly around a specified average.

Rational emotive therapy A Cognitive Therapy based on Albert Ellis' theory that cognitions control our emotions and behaviours; therefore, changing the way we think about things will affect the way we feel and the way we behave.

Rationalization In psychoanalytic theory, a mechanism of defence by means of which unacceptable thoughts or impulses are reinterpreted in more acceptable and, thus, less anxiety-arousing terms (e.g., the jilted lover who convinces herself that she never loved her fiancé anyway).

Raven's progressive matrices test An IQ test that gauges the subject's ability to solve problems that are presented in unfamiliar designs.

Raw data The initial data gathered that has not yet been graphed, organized, or analyzed.

R-data A category of data used in personality assessment, relating to others' ratings of a target individual.

Reactance 1. A motive to protect or restore one's sense of freedom. Reactance arises when someone threatens our freedom of action.

2. The desire to assert one's sense of freedom.

Reaction formation A defence mechanism that involves behaving in a way that is the opposite of how one really feels because the true feelings produce anxiety.

Reaction range The limits that heredity places on characteristics such as IQ.

Reaction time The amount of time a subject takes to respond to a stimulus.

Reactive effects of experimental arrangements If the experimental setting is not the same as the natural setting that the participants would usually be in, external validity may be threatened by the participants reactions to the different environment, I.e. results can not be generalized to the natural setting.

Reactive effects of testing/pretest sensitization The pretest may influence the participant's response to experimental treatment. Also, the general population is not receiving the pre-test, and so the sample response cannot necessarily be generalized to the rest of the population.

Reactive schizophrenia According to Bleuler, a form of schizophrenia characterized by rapid onset and brief duration; he assumed the cause was stressful life situations.

Readiness tests Tests to assess the student1s level of skills and knowledge necessary for a given activity.

Readiness training Teaching the skills and knowledge necessary for a given activity.

Reading Recovery A program that provides one-to-one tutoring from specially trained teachers to first-graders who are not reading adequately.

Realistic group conflict theory The theory that prejudice arises from competition between groups for scarce resources.

Reality principle In Freud's theory, the set of rules that govern the ego and that dictate the way in which it tries to satisfy the id by gaining pleasure in accordance with the real world and its demands.

Reality testing The knack of distinguishing between inner and outer, fantasy and externality.

Reasoning The determination of the conclusions that can be drawn from certain premises.

Recall A method of retrieval in which an individual is required to reproduce the information previously presented.

Recasting The often instinctive way in which the young child's ungrammatical utterances are put into correct grammatical forms by its mother or carer and which aid in the development of the child's language.

Receiver operating characteristic curve (ROC curve) A graph of hits and false alarms of subjects under different motivational conditions; indicates people's ability to detect a particular stimulus.

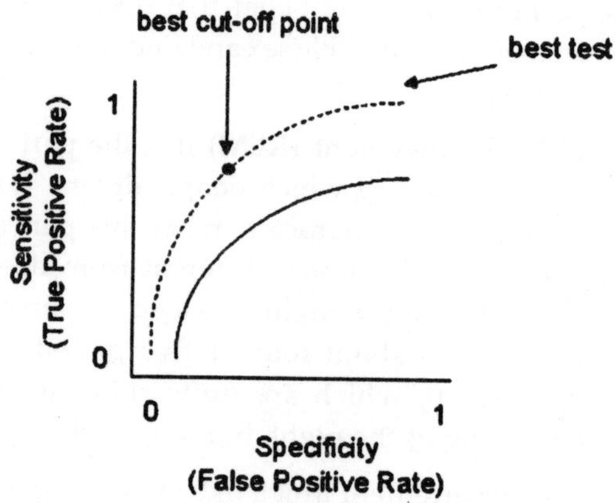

Fig. ROC curve

Recency effect The tendency to recall later information. In the memorization of a list of words, the recency effect is evidenced by better recall of the last words in the list.

Receptive field That portion of the visual field in which the presentation of visual stimuli will produce an alteration in the firing rate of a particular neuron.

Receptive language An understanding of the spoken and written word.

Receptor cells A special type of neuron that can respond to various external energies and transduce (translate) physical stimuli into electrical changes to which other neurons can respond.

Receptor molecule A special protein molecule located in the membrane of the postsynaptic neuron that responds to molecules of the transmitter substance. Receptors such as those that respond to opiates are sometimes found elsewhere on the surface of neurons.

Recessive allele The form of the gene that does not influence the expression of a trait unless it is paired with another recessive allele.

Reciprocal altruism Altruism in which people behave altruistically toward one another because they are confident that such acts will be reciprocated toward either them or their kin.

Reciprocal determinism A concept of Albert Bandura's sociallearning theory that refers to the notion that a complex reciprocal interaction exists among the individual, his or her behaviour, and environmental stimuli and that each of these components affects the others.

Reciprocal helping (Also know as reciprocal altruism.) A sociobiological principle stating that people expect that anyone helping another will have that favour returned at some future time.

Reciprocal inhibition The arrangement by which excitation of some neural system is accompanied by inhibition of that system's antagonist (as in antagonistic muscles).

Reciprocal interaction The fact that different people seek out different situations.

Reciprocal teaching A teaching method based on the principles of question generation, in which metacognitive skills are taught through instruction and teacher modeling to improve the reading performance of students who have poor comprehension.

Reciprocity A mutual exchange between what we give and receive—for example, liking those who like us.

Reciprocity norm Expectation that favours will be returned – if someone does something for another person, that person should do something in return.

Reciprocity principle A basic rule of many social interactions that decrees that one must repay whatever one has been given.

Recoding Changing the form in which information is stored.

Recognition A method of retrieval in which an individual is required to identify stimuli as having been experienced before.

Recognition A task in which a participant must judge whether he has encountered a stimulus previously.

Recognition by components A theory of visual object recognition, holding that objects are represented internally as a constellation of geons and that the image is segmented into the parts corresponding to geons so that the resulting description can be matched to the appropriate object description.

Reconditioning In classical conditioning, the presentation of further reinforced conditioning trials (i.e., those that include the unconditioned stimulus [US]) after a conditioned response (CR) has been extinguished.

Reconstruction Tendency to fill in the gaps in our memory and often believe these represent true memories.

Reconstructive memory The process of putting information together based on general types of stored knowledge in the absence of a specific memory representation.

Rectophobia Fear of rectum or rectal diseases.

Reductionist psychology A form of psychology which is based upon an examination of phenomena through the study of their individual components; thus, in social psychology, the study of the group through the psychology of the individuals that make up that group.

Reductionistic perspective Studies human development by reducing complex phenomenon or events to a single cause.

Reference An extra-linguistic definition of meaning, also known as extension, where the reference of an expression is what entities or states it points to in the outside world.

Reference group A group to which people orient themselves, using its standards to judge themselves and the world.

Referent power Power given to an individual due to respect and/or desire to be similar to that individual.

Reflectance The proportion of light aimed at an object that is reflected by it. An object's reflectance determines whether the object is perceived as light or dark.

Reflected appraisals In symbolic interactionism, the way in which others actually evaluate us when mediated by how we believe others evaluate us, a fact which is strongly related to our self-concepts. Also known as looking-glass self.

Reflection A therapeutic technique in humanistic therapy where the feelings and thoughts of the client are reflected or reworded back to the client to assist in understanding them.

Reflectivity The act of analyzing oneself and one's own thoughts.

Reflex A particular pattern of behaviour, such as sucking, triggered by a specific stimulus with which all children are born and which plays a crucial part in the overall development of the child in the months before and after birth.

Refractory periods The waiting time before men can regain an erection after they have ejaculated.

Regression A return, temporary or chronic, to an earlier level of psychological development. (An example would be the second Bush administration.).

Regression analysis Used with a correlation to determine a regression equation that predicts or estimates a persons score on one variable if the other is known.

Regression toward the average The statistical tendency for extreme scores or extreme behaviour to return toward one's average.

Regression toward the mean The tendency for extreme states to move toward the average when assessed a second time.

Regrouping A method of ability grouping in which students in mixed-ability classes are assigned to reading or math classes on the basis of their performance levels.

Regulatory behaviour A behaviour that tends to bring physiological conditions back to normal, thus restoring the condition of homeostasis.

Rehearsal Mental repetition of information, which can improve its retention.

Reinforced trial In classical conditioning, a trial on which the conditioned stimulus (CS) is followed by the unconditioned stimulus (US). In instrumental conditioning, a trial in which the instrumental response is followed by reward, cessation of punishment, or some other reinforcer.

Reinforcement In classical conditioning, the procedure by which the unconditional stimulus (US) is presented contingent upon the occurrence of the conditioned stimulus (CS). In instrumental conditioning, the procedure by which a sought-after consequence is made contingent upon the occurrence of the instrumental response.

Reinforcement contingency A consistent relationship between a response and the changes in the environment that it produces.

Reinforcement schedule The pattern in which reinforcement is given over time.

Reinforcer Event or thing added to or removed from a situation that is contingent on the occurrence of a behaviour. Probability of occurrence (POO) or the related behaviour remains unchanged or increases.

Relational aggression A strategy for attaining social advantage by manipulating others' social alliances. Females are apparently more relationally aggressive, whereas males are apparently more physically aggressive.

Relative deprivation Feelings of discontent aroused by the belief that one fares poorly compared with others.

Relative grading standard Grading on the basis of how well other students performed on the same test rather than in terms of preestablished absolute standards.

Relative motion parallax A source of information about depth in which the relative distances of objects from a viewer determine the amount and direction of their relative motion in the retinal image.

Relative size A monocular depth cue in which far-off objects produce a smaller retinal image than nearby objects of the same size.

Relaxation procedures Procedures by which a person can made to relax.

Relaxation response A condition in which muscle tension, cortical activity, heart rate, and blood pressure decrease and breathing slows.

Relearning A method for measuring forgetting and retention, which involves assessing the amount of time it takes to memorize information a second time.

Releasing stimulus A term used by ethologists to describe a stimulus that is genetically programmed to elicit a fixed-action pattern (e.g., a long, thin, red-tipped beak which elicits a herring gull chick's begging response).

Reliability The degree of consistency with which a test measures a trait or attribute. Assuming that a trait or attribute remains constant, a perfectly reliable test of that measure will produce the same score each time it is given.

Reliability coefficient The coefficient used in determining the consistency of mental tests, that is, the repeatability of their results. It is usually derived from test-retest correlations or from correlations between alternative forms of a test.

Reliable That which can be relied on; dependable; trustworthy; in testing, provides consistent results when administered on different occasions.

Religion A childlike yearning for an all-powerful Parent to take away feelings of helplessness that arise from confronting the forces of nature. A collective neurosis. An illusion (not necessarily an error) arising from childish wishes, religion spares many a believer an individual neurosis by reducing him to "psychical infantilism." The first religion was a form of totemism.

Reluctant witness Witness responds with reluctance due to a natural restraint that is a reflection of their personality. This particular witness will be hesitant and reserved. They will not find it easy to talk freely. Others may feel that the act they have witnessed is "none of their business".

REM rebound The tendency to spend more time in REM sleep if deprived of it on previous nights. REM rebound often occurs during withdrawal from medications that suppress REM sleep (e.g., barbiturates or alcohol).

REM sleep behaviour disorder A neurological disorder characterized by absence of the paralysis that normally occurs during REM sleep; the patient acts out his or her dreams.

REM sleep The type of sleep characterized by rapid eye movements, an EEG indicative of high cortical arousal, speeded heart rate and respiration, near-paralysis of limb muscles, and recall of highly visual dreams.

Remediation Instruction given to students having difficulty learning.

Removal punishment Decreasing the chances that a behaviour will occur again by removing a pleasant stimulus following the behaviour.

Repertory grid In psychotherapy, a grid method, base on an individual's particular judgements about people in their environment and from which constructs can be constructed to be elaborated on in interview.

Repetition compulsion The urge to do the same thing over and over, especially in terms of relationship patterns (e.g., always winding up with the same kind of partner). Rather than seeing it as unlearned lessons reconstellated in the present, Freud linked this to the death drive's push toward a conservative return to previous conditions.

Repetition priming An increase in the likelihood that an item will be identified, recognized, or recalled caused by recent exposure to that item, which may occur without explicit awareness.

Replicability The ability of research to repeatedly yield the same results when done by different researchers.

Replication The strength of a research study is only as good as its ability to be replicated. In other words, if a study has significant results but can not be done again, it is difficult to assess whether it was a good study or a result of error.

Representational thought In Piaget's theory, thought that is internalized and includes mental representations of prior experiences with objects and events.

Representational understanding Also referred to as insight, the capacity to understand how things work and thereby choose or modify objects that meet precise criteria for tool use, in advance of undertaking a task.

Representations Cognitions that correspond to (represent) certain events, or relations between events, in the world.

Representative sample A subset of a population that closely matches the overall characteristics of the population with respect to the distribution of males and females, racial and ethnic groups, and so on.

Representativeness One of three types of heuristic, according to which a person, thing or event will be judged to be a member of a class whose stereotypical members it closely resembles, regardless of other information.

Representativeness heuristic A cognitive strategy that assigns an object to a category on the basis of a few characteristics regarded as representative of that category.

Repressed memory In psychoanalytic theory, a memory that is so anxiety-laden that it has been pushed out of consciousness where it may fester until it is "recovered." There is little scientific evidence for the existence of recoverable repressed memories.

Repression The ego's ridding itself of unacceptable desires and ideas by dumping them into unconsciousness. Three conditions lead to a return of the repressed into consciousness: when the counter-cathexis holding it down weaken; when the drives below it get stronger; or when recent events similar in theme to what's repressed reawaken it; in all three cases the repressed shows up in the very method used for repression (e.g., an archeologist who represses his erotic nature falling in love with an alluring statue). "Primal repressions" are handed down from the painful experiences of prior generations (example: incest taboo) and keep certain contents of the id from ever becoming conscious. "Repression proper" (also called "after-expulsion") has to do with the repression of associations and ideas related to whatever idea is chiefly being repressed. An aim of psychoanalysis is to turn repression into condemnation.

Note: in German, Freud wrote about verdrangung, which really ought to be translated into.

Reproductive advantage The outcome of a characteristic that helps an organism mate successfully and thus pass on its genes to the next generation.

Reproductive strategies Different systems of mating and rearing offspring. These include monogamy, polygny, polyandry, and polygamy.

Reproductive success The number of viable offspring an individual produces relative to the number of viable offspring produced by other members of the same species.

Research hypothesis A statement which indicates that there is an expected relationship between the independent and dependent variables.

Residual stress pattern A chronic syndrome in which the emotional responses of posttraumatic stress persist over time.

Resistance The patient's effort to remain unconscious of what is repressed. Common forms of resistance include arguing with the therapist, refusing to free associate, forgetting to show up for a session, reducing spirit to a compensation for helplessness, and deploying defence mechanisms in order to remain unaware of unconscious material. The negative therapeutic reaction is the most difficult form of resistance and is motivated by the need to expiate an unconscious sense of guilt via self-defeating behaviour.

Resistant attachment A kind of attachment in which infants show mixed reactions to their mothers. They may approach their mothers upon their return but, at the same time, continue to cry or even push their mothers away.

Resource dilemma Social dilemmas concerning how two or more people share a limited resource.

Response Behaviour or group of behaviours to be analyzed – target behaviour.

Response amplitude The size of a response, used commonly as a sign of response strength in classical and operant conditioning.

Response bias In signal detection experiments, the willingness of the participant to assert that a stimulus has occurred or not occurred, given the costs of false positives or false negatives.

Response cost A consequence that decreases the frequency of response that is regularly and reliably followed by the termination of an appetitive stimulus.

Response latency The time that elapses between an elicitor and a response (in classical conditioning, between the onset of a conditioned stimulus [CS] and the conditioned response [CR]). As classical conditioning proceeds, response latency usually decreases (i.e., the CR occurs sooner).

Response prevention A therapeutic technique where stimuli is presented to the client but the client is not permitted to exercise his or her typical response. Used for the treatment of phobias, obsessive compulsive disorder and other anxiety disorders.

Response rate The number of responses per unit of time. This is one measure of the strength of an operant response.

Response suppression The inhibition of a response by conditioned fear.

Response tendency A learned tendency to behave in a particular way.

Resting potential The polarization of cellular fluid within a neuron, which provides the capability to produce an action potential.

Resting state The period during which the inside of a neuron has a slightly higher concentration of negatively charged ions than the outside does. A neuron during this time is inactive.

Restitutional symptoms For Eugen Bleuler, symptoms such as delusions and hallucinations that originated in the schizophrenic patient's attempt to compensate for his increasing isolation from the world.

Restrained-eating hypothesis The hypothesis that obese persons are oversensitive to external food cues because their conscious restraints on eating have become disinhibited.

Restructuring A reorganization of a problem that can facilitate its solution; a characteristic of creative thought.

Retention The proportion of learned information that is retained or remembered.

Retention intervals In memory experiments, the time that elapses between the original learning and a later test.

Reticular formation The region of the brain stem that alerts the cerebral cortex to incoming sensory signals and is responsible for maintaining consciousness and awakening from sleep.

Retina The tissue-thin structure at the back of the interior of the eye that contains the photoreceptors, several layers of intermediate neurons, and the cell bodies of the axons that form the optic nerve.

Retinal disparity The binocular cue to distance referring to the distance between the two images sent to the brain by our eyes. The farther apart these images, the closer the object.

Retinal ganglion cells The output stage of the retina and the first stage of processing at which action potentials are generated. Retinal ganglion cells have characteristic circular.

Retinal image The image of an object that is projected on the retina. Its size increases with the size of that object and decreases with the object's distance from the eye.

Retinal receptor The light sensitive cells in the retina where light is converted into a neural response. In human beings this is comprised of a mixture of rods and cones.

Retinex theory A theory of colour vision proposing that the amount of light in the image is first analysed separately at different wavelengths and that the resulting maps of the image are subsequently compared.

Retinotopic map The orderly projection of neurones from the retina so that the spatial structure of the image is preserved at the destination, for example, nearby cells in the primary visual cortex have receptive fields in nearby regions of the image.

Retrieval The process of searching for some item in memory and of finding it. If retrieval fails, this may or may not mean that the relevant memory trace is missing. The trace may simply be inaccessible.

Retrieval cues Contextual variables, including physical objects, suggestions, or other verbal stimuli, that improve the ability to recall information from memory.

Retrieval failure The inability to access a memory, often due to poor encoding; an alternative to erasure as an explanation for forgetting.

Retroactive facilitation Increased comprehension of previously learned information due to the acquisition of new information.

Retroactive inhibition Decreased ability to recall previously learned information causedby learning of new information.

Retroactive interference Forgetting of old information because of newly learned information.

Retrograde amnesia A memory deficit, often suffered after a head injury, in which the patient loses memory of some period prior to the injury.

Retrospective study A research technique that requires subjects to report what happened in the past.

Reuptake The process by which a terminal button retrieves the molecules of transmitter substance that it has just released; terminates the effect of the transmitter substance on the receptors of the postsynaptic neuron.

Reversal design Any single subject design that includes the removal of treatment to determine if the subject reverts to baseline (ex. ABA, ABAB).

Reversal theory Theory that explains human motivation in terms of reversals from one to the other opposing metamotivational states.

Reversibility A child's ability to reverse operations and therefore recognize that the qualities of an object remain the same despite changes in appearance. Occurs in Piaget's Concrete Operational Stage of Cognitive Development (e.g., 1+2=3 to 3-2=1).

Reversible figures Visual patterns that allow parsing such that what is initially figure becomes ground and vice versa (e.g., a drawing that can be seen as either a black picket fence against a white wall or a white picket fence against a black wall).

Reward power Power derived through an ability to offer rewards.

Reward theory of attraction The theory that we like those whose behaviour is rewarding to us or whom we associate with rewarding events.

Rh factor Incompatibility group of antigens, determined by heredity and usually present in human red blood cells, that may cause hemolytic reactions during pregnancy or after transfusion of blood containing this factor into someone lacking it.

Rhabdophobia Fear of being severely punished or beaten by a rod, or of being severely criticized. Also fear of magic.(wand).

Rhodopsin The photopigment used in the rods within the retina.

Rhypophobia Fear of defecation.

Rhytiphobia Fear of getting wrinkles.

Risk aversion In decision making, the reluctance to choose an alternative that involves some risk.

Risk factor The element of risk involved in any exposure to a potentially harmful agent, for example that of the developing foetus to developmental abnormality, through the use by the mother of harmful drugs, cigarettes, alcohol during pregnancy.

Risk taking A form of conscious human behaviour influenced by a calculation of risk against possible achievement or reward to be gained, which reaches its peak amongst young men during later adolescence.

Risky shift The tendency for a dominant, risky point of view in a group to be strengthened to an even riskier position after a group discussion.

Rites of passage Ceremonies employed by cultures to mark important developmental transitions (e.g., puberty rites, debutante balls, graduation exercises).

Ritual healing Ceremonies that infuse special emotional intensity and meaning into the healing process.

Robber's cave A term for the study of group behaviour among children, based on the work of muzafer and carolyn sherif, who set up a classic series of natural experiments at summer camps in the 1950s where they identified the strengths and conflicts of a miniature society and set about resolving the conflicts that had been created.

Rods Photoreceptors concentrated in the periphery of the retina that are most active in dim illumination; rods do not produce sensation of colour.

Rogers, Carl A humanistic Psychologist who developed Client-Centered Therapy.

Role A set of norms that define how people in a given social position ought to behave.

Role playing In psychotherapy, a technique in which the therapist and patient act out scenes, such as a marital confrontation, that are likely to occur outside the therapeutic situation.

Roles and memberships In the system for analyzing responses to the 'who am I?' devised by C. Gordon, those groups that a person has had some choice about joining, as opposed to the ascribed characteristics assigned to them at birth.

Roles and specialisms In group membership, the skills manifested by certain members as leaders or experts within the group and whose role is thus more social- than task-related.

Romantic love A state of emotion characterized by idealization of the beloved, turbulent emotions, and obsessive thoughts.

Romeo-and-Juliet effect The intensification of romantic love that can occur with parental opposition.

Rooting reflex In the infant, the sucking elicited by stroking applied on or around the lips; aids breast-feeding.

Rorschach inkblot test A projective (unstructured) technique of personality assessment that requires the examinee to look at a series of inkblots and report everything she sees in them.

Rote learning Memorization of facts or associations.

Rough-and-tumble play A type of play amongst children, involving chasing and tackling each other, which most children by the age of 8 are able to discern from aggressive fighting.

Round window An opening in the bone surrounding the cochlea. Movements of the membrane behind this opening permit vibrations to be transrnitted through the oval window into the cochlea.

Rubella A mild, infectious, communicable virus disease, characterized by swollen glands, especially of the back of the head and neck, and small red spots on the skin; German measles.

Rule example rule Pattern of teaching concepts by presenting a rule or definition, giving examples, and then showing how examples illustrate the rule.

Rules Behavioural guidelines for acting in certain ways in certain situations.

Rules of inference A set of theories applied in the study of how people reason, the best known of which is the modus ponens, and which suggests that people encode what they know as sentences with certain formal structures and use a set of mentally encoded rules of inference to see what other sentences can be derived from them.

Rupophobia Fear of dirt.

Russophobia Fear of Russians.

S

S factor According to Spearman, a factor of intelligence that is specific to a particular task.

Saccadic movement The rapid movement of the eyes that is used in scanning a visual scene, as opposed to the smooth pursuit movements used to follow a moving object.

Sadistic, sadism A person whose defence is channellising one's frustration into aggressive acts against others. A sadist is a person who finds pleasure in hurting others or seeing others in pain.

Safety signal A cue which predicts that an aversive stimulus will not occur.

Sagittal section A sectional cut across the brain made vertically from front to back.

Samhainophobia Fear of Halloween.

Sample A subset of a population selected by the investigator for study. A random sample is constructed such that each member of the population has an equal chance of being picked. A stratified sample is constructed such that every relevant subgroup of the population is randomly sampled in proportion to its size.

Sample mean Abbreviated with a lowercase x with a horizontal line over top (called 'x-bar'), the sample mean is the true mean of a sample of data often used to estimate the true mean of the entire population.

Sample size The more heterogeneous a population, the larger the sample needs to be. A large sample is needed when looking for small differences.

Sample standard deviation The standard deviation of a sample of the population. Often used to estimate the true population standard deviation. Often abbreviated 'SD.".

Sampling bias A source of error that arises when the sample is not representative of the population that the researcher wants to study.

Sampling distribution of means The distribution of the means from samples drawn from a population.

Sampling error The amount of error associated with a sample due to its deviation from the population.

Sankhya yoga System of philosophy and yoga (main concepts - purusha and prakriti) of Kapila (7th century B.C.).

Sapir-Whorf hypothesis An hypothesis associated with Benjamin Whorf and Edward Sapir that there must be an intimate connection between human language and human thinking. It is often presented in two versions: in which language determines thought (linguistic determinism); or in which different languages will encode different thoughts (linguistic relativity).

Sarmassophobia Fear of love play. (Malaxophobia).

Satanophobia Fear of Satan.

Satsang Mixing with or keeping the company of people having high Stability.

Saturation The intensity of a colour, for example, pink and red have the same hue, but red is more saturated. The most saturated colours are produced by monochromatic light.

Savant A mentally retarded person who has some remarkable talent that seems out of keeping with his low level of general intelligence. Previously *idiot savant,* a term now abandoned as derogatory.

Scabiophobia Fear of scabies.

Scaffold The constructive support system provided by mothers and carers in order for the child to progress through the stages of language development.

Scaffolding Support for learning and problem solving. The support could be clues, reminders, encouragement, breaking the problem down into steps, providing an example, or anything else that allows the student to grow in independence as a learner.

Scaling A procedure for assigning numbers to a subject's responses.

Scalloped response pattern The phenomenon in which responses are slow in the beginning of the interval and faster just before reinforcement happens. It occurs as a result of a fixed-interval schedule.

Scatophobia Fear of fecal matter.

Scatter plot A graph depicting the relationship between two interval- or ratio-scale variables, with each axis representing one variable; often used to graph correlation data.

Scelerophibia Fear of bad men, burglars.

Schedule of reinforcement The pattern of occasions on which responses are to be reinforced. Commonly, reinforcement is scheduled after a stipulated number of responses occurs or when a response occurs after a preset time interval has elpased.

Schema 1. In theories of memory and thinking, a term that refers to a general cognitive structure in which information is organized.

2. In Piaget's theory of development, a mental pattern.

Schema theory Theory that information is stored in long-term memory in networks of connected facts and concepts that provide a structure for making sense of new information.

Schemata Mental networks of related concepts that influence understanding of new information.

Schematics In experiments, those people who rate themselves highly in a particular domain and who see that domain as important to them.

Schemes Piaget's term for cognitive structures that develop as infants and young children learn to interpret the world and adapt to their environment.

Schizoid personality disorder A disorder characterized by social withdrawal and restricted expression of emotions.

Schizophrenia A group of severe mental disorders characterized by at least some of the following: marked disturbance of thought, withdrawal, inappropriate or flat emotions, delusions, and hallucinations. *See also* catatonic schizophrenia, disorganized type of schizophrenia, negative symptoms of schizophrenia, paranoid schizophrenia, positive symptoms of schizophrenia.

Scientific jury selection A method of selecting juries through surveys that yield correlations between demographics and trial-relevant attitudes.

Scientific method A framework of approaches and procedures for forming a hypothesis and gathering and interpreting objective information through experimentation. Seeks to minimize sources of bias and to yield dependable, and independently testable, information.

Sciophobia sciaphobia Fear of shadows.

Sclera The tough outer layer of the eye; the "white" of the eye.

Scoleciphobia Fear of worms.

Scolionophobia Fear of school.

Scopophilia The pathological enjoyment that comes with watching sexual activity. This is a mistranslation of Schaulust, "sexual pleasure in looking.".

Scopophobia or scoptophobia Fear of being seen or stared at.

Scotoma In cases of partial blindness, a neglect or blindness in the retina for a certain part of the visual field.

Scotomaphobia Fear of blindness in visual field.

Scotophobia Fear of darkness. (Achluophobia).

Scotopic vision Nighttime vision, mediated primarily by rods and therefore without colour information.

Screen memories Memory fragments that cover a child's earliest and forgotten sexual fantasies.

Script A representation or scenario which captures a sequence of events that usually happen when people meet.

Scriptophobia Fear of writing in public.

Search for cheaters In social contract theory, an instruction to focus on instances of cheating, not just where only one side cheats, but also where both sides can cheat.

Seasonal affective disorder A mood disorder characterized by depression, lethargy, sleep disturbances, and craving for carbohydrates. This disorder generally occurs during the winter, when the amount of daylight, relative to the other seasons, is low. This disorder can be treated with exposure to bright lights.

Seasonal affective disorder A mood disorder that shows reliable fluctuations with the time of year. One example is a depression that ensues in the fall when the days become shorter and ends in the spring when the days lengthen.

Seasons A metaphor, applied by daniel levinson to his concept and description of the human developmental eras, where life is seen in the familiar, poetic terms of e.g. Spring, summer, autumn, winter.

Seatwork Work that students are assigned to do independently during class.

Second messengers Neurochemicals within the neuron that regulate such mechanisms as the creation of receptors sites for specific neurotransmitters and the synthesis of the neuron's own neurotransmitter, thus determining the neuron's overall responsiveness.

Secondary appraisal In the study of psychological stress, the second of two processes applied by the individual when they appraise their ability to cope with a particular threat.

Secondary elaboration The dreamer's tendency to fill in gaps in the dream and explain its mysteries in order to understand it better. This usually leads to significant misunderstandings of the dream's true latent content.

Secondary process The ego's reality-testing and energy-binding capability. In other words, thinking and reasoning.

Secondary process thinking Thinking that is logical and rational.

Secondary punisher In operant conditioning, a consequence that is unpleasant because it has become associated with a primary punisher. It is also called a conditioned punisher.

Secondary reinforcer In operant conditioning, a consequence that is satisfying because it has become associated with a primary reinforcer. It is also called a *conditioned reinforcer.*

Secondary sexual characteristic Any of the physical characteristics that differentiate male and female individuals, as distribution of hair or fat on the body, breast and muscle development, deepening of the voice, and so on, that are not directly related to reproduction and usually appear at puberty.

Secure attachment A kind of attachment in which infants use their mothers as a base for exploring a new environment. They will venture out from their mothers to explore a Strange Situation but return periodically.

Secure attachment style An expectation about social relationships characterized by trust, a lack of concern with being abandoned, and a feeling of being valued and well liked.

Sedatives Drugs that slow down the nervous system.

Selachophobia Fear of sharks.

Selaphobia Fear of light flashes.

Selection A threat to internal validity caused by the non-random assignment of subjects to groups. When subjects are not assigned at random, there is a strong possibility that the two groups are not initially the same in certain important respects.

Selection bias Errors in the selection and placement of subjects into groups that results in differences between groups which could effect the results of an experiment.

Selective attention The process that controls our awareness of, and readiness to respond to, particular categories of stimuli or stimuli in a particular location.

Selective optimization A strategy for fulfillment throughout the aging process, where one maximizes gains and minimizes losses associated with growing older. In other words, making the best of what you have.

Selective optimization with compensation A strategy for successful aging in which one makes the most of gains while minimizing the impact of losses that accompany normal aging.

Selective recall A phenomenon of extremely detailed memory recall when a suspect is asked to relate his whereabouts/actions during the offence. This airtight and precise recall will not reflect similar recall of other time periods preceding or following the offence.

Selective serotonin reuptake inhibitors (ssris) Medications such as Prozac, Zoloft, and Paxil that increase serotonin turnover in the brain and find wide use as treatments for depression, obsessive-compulsive disorder, panic disorder, and many other disorders.

Selective social interaction theory The view that suggests that, as people age, they become more selective in choosing social partners who satisfy their emotional needs.

Selenophobia Fear of the moon.

Self A symbol-using individual who can reflect upon his/her own behaviour.

Self-actualization Term used by Goldstein and Maslow to denote actualization one's potentialities.

Self-affirmation theory A theory predicting that people will often cope with specific threats to their self-esteem by reminding themselves of other unrelated but cherished aspects of their self-concept.

Self-analysis Freud conducted a long analysis of himself during the time he formulated the first psychoanalytic concepts. He believed much could be learned by analyzing oneself, but that being analyzed by someone else was necessary due to the many resistances and blind spots that occur in the course of self-observation. Asked about the number of analytic pioneers who had not been analyzed, he replied that they definitely weren't the better for it.

Self-attribution Attributions made about the causes of our behaviour based on our self-observations of the way we act in different situations.

Self-authentication An attempt to learn the truth about oneself by carrying out limited social experiments.

Self-awareness A self-conscious state in which attention focuses on oneself. It makes people more sensitive to their own attitudes and dispositions.

Self-awareness theory The theory that self-focused attention leads people to notice self-discrepancies, thereby motivating either an escape from self-awareness or a change in behaviour.

Self-complexity The number of distinct roles or identities people believe they have.

Self-concept According to Rogers, the most important feature of personality. The self-concept includes all the thoughts, feelings, and beliefs people have about themselves.

Self-confidence Freedom from doubt, belief in yourself and your abilities.

Self-consciousness The habitual tendency to engage in self-awareness.

Self-consistency The validation of the self by confirming one's beliefs about the self.

Self-control Behaviour that produces the larger, longer-term reward when people are faced with the choice between it and the smaller, short-term reward.

Self-disclosure Revealing intimate aspects of oneself to others.

Self-discrepancy theory A theory that people experience specific negative emotions when they perceive a discrepancy between their self-concept and various self-guides.

Self-effacing bias The tendency of people in certain cultures to attribute their successes to situational factors rather than to personal attributes and to attribute their failures to lack of effort.

Self-efficacy A sense that one is competent and effective, distinguished from self-esteem, one's sense of self-worth. A bombardier might feel high self-efficacy and low self-esteem.

Self-efficacy theory A theory that motivation is determined both by the belief that one is capable of successfully performing some behaviour, and by the belief that performing the behaviour will lead to certain outcomes.

Self-enhancement The process of seeking out and interpreting situations so as to attain a positive view of oneself.

Self-esteem A generalized evaluative attitude toward the self that influences both moods and behaviour and that exerts a powerful effect on a range of personal and social behaviours.

Self-evaluation maintenance model A theory predicting under what conditions people are likely to react to the success of others with either pride or jealousy.

Self-fulfilling prophecy The process by which one's expectations about a person eventually lead that person to behave in ways that confirm those expectations.

Self-handicapping A self-protective strategy in which one arranges for an obvious, and nonthreatening obstacle to one's own performance, such that any failure can be attributed to the obstacle and not to one's own limitations.

Self-help groups Groups that are similar to therapy groups except that they do not have a therapist.

Self-hypnosis The act or process of hypnotizing oneself.

Self-identity In the theory of reasoned action, a determinant of behaviour that is just as important as attitude in predicting a particular action.

Self-monitoring The process of making sure that one's own behaviour conforms with the demands of the current social situation. At the extremes, high self-monitors are pliant social chameleons, and low self-monitors are rigid and unaccommodating.

Self-perception theory The idea that people observe themselves in order to figure out the reasons they act as they do; people infer what their internal states are by perceiving how they are acting in a given situation.

Self-perception theory The theory that when internal cues are difficult to interpret, people gain self-insight by observing their own behaviour.

Self-presentation The act of expressing oneself and behaving in ways designed to create a favourable impression or an impression that corresponds to one's ideals.

Self-promotion Conveying positive information about the self either through one's behaviour or by telling others about one's positive assets and accomplishments.

Self-reference effect The tendency to process efficiently and remember well information related to oneself.

Self-regulated learners Students who have knowledge of effective learning strategies and how and when to use them.

Self-regulation The ability to think and solve problems without the help of others.

Self-regulatory systems For Walter Mischel, ways that individuals regulate their own behaviour using self-imposed goals and plans.

Self-report The result of a psychological experiment contributed by a person involved which throws light on the quality of their experience, usually in the form of an interview or survey, a test, introspection, think-aloud protocol, or even a diary study.

Self-report data Information that people being surveyed give about themselves.

Self-report inventory A paper-and-pen test that requires people to answer questions about their typical behaviour.

Self-report measures The self-behaviours that are identified through a participant's own observations and reports.

Self-schema A mental framework that represents and synthesizes information about oneself; a cognitive structure that organizes the knowledge, feelings, and ideas that constitute the self-concept.

Self-serving attributional bias The tendency, found largely in individualistic cultures, to deny responsibility for failures but take credit for successes.

Self-serving bias In the study of the attribution process, the tendency to attribute one's success to internal factors and one's failure to external factors, thus avoiding blame for negative events and accepting credit for positive events, which in turn increase our self-esteem.

Self-socialization The process whereby a child develops a sense of gender identity in the course of sex-role identification, and which is not directly dependent on external reinforcement.

Self-verification The process of seeking out and interpreting situations so as to confirm one's self-concept.

Semantic encoding A way of encoding verbal information that emphasizes the meaning of words.

Semantic feature The smallest significant unit of meaning within a word (e.g., male, human, and adult are semantic features of the word *man*).

Semantic grammar A grammar which seeks to directly deliver the meaning of any sentence, before (or without regard to) its syntax.

Semantic memories Generic, categorical memories, such as the meanings of words and concepts.

Semantic priming A facilitating effect on the recognition of words having meanings related to a word that was presented previously.

Semantic slanting A way of making statements so that they will evoke specific emotional responses.

Semantics The meanings and the study of the meanings represented by words.

Semi interquartile range One half of the interquartile range.

Semicircular canals The three curved tubules found within the inner ear that contain a viscous liquid that is easily perturbed if jostled. The canals provide moment-to-moment information about head movements.

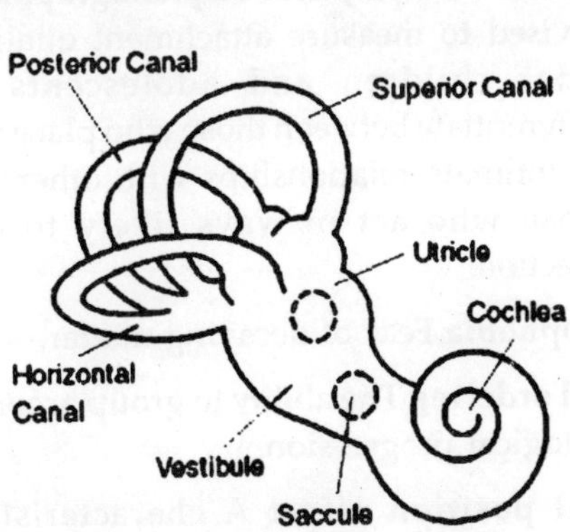

Fig. Semicircular canals

Semistructured interview A technique developed by james marcia which put the ideas of erik erikson on adolescent development into a testable form.

Senile Showing the marked deterioration often accompanying old age, especially mental impairment characterized by confusion, memory loss, and so on.

Sensation The power or process of receiving conscious sense impressions through direct stimulation of the bodily organism; an immediate reaction to external stimulation of a sense organ; conscious feeling or sense impression.

Sensation seeking A predisposition to seek novel experiences, look for thrills and adventure, and be highly susceptible to boredom.

Sense Also known as intension this is a linguistic definition of meaning, where an expression can be defined by purely linguistic things.

Sensorimotor Of or pertaining to motor responses initiated by sensory stimulation.

Sensorimotor communication The baby's early signalling system, which emphasizes the fact that communication in the first year and a half is rooted to vocalizations and gestural actions centred around feeding, dressing, bathing and play.

Sensorimotor period The first period in Piaget's theory of cognitive development, from birth to 2 years. Marked by an orderly progression of increasingly complex cognitive development: reflexes, permanence, appreciation of causality, imitation, and symbolic thinking.

Sensorimotor stage The first stage in Piaget's Stages of Cognitive Development where a child's primary way of learning about the world is through the senses and movement.

Sensory adaptation A phenomenon in which receptor cells lose their power to respond after a period of unchanged stimulation;

allows a more rapid reaction to new sources of information.

Sensory association cortex Those regions of cerebral cortex that receive information from the primary sensory areas.

Sensory code The rule by which the nervous system represents the sensory characteristics of the stimulus. One example is firing frequency which, in touch and vision, encodes increased stimulus intensity.

Sensory coding The process by which the sensory organs translate various stimulus qualities into the dimensions of our sensory experience.

Sensory impairments Problems with the ability to receive information through the body1s senses.

Sensory interaction The increased or decreased sensitivity of the sensory system in response to other ongoing stimulation.

Sensory memory A memory system that stores incoming sensory information for an instant.

Sensory modalities A technical term for the sensory domains: taste, touch, smell, kinesthesis, vision, and hearing.

Sensory neuron A neuron that detects changes in the external or internal environment and sends information about these changes to the central nervous system.

Sensory organs Specialized structures of the body containing sensory receptors that receive stimuli from the environment.

Sensory physiology The study of the way in which biological mechanisms convert physical events into neural events.

Sensory process According to signal detection theorists, the underlying neural activity in the sensory system upon which all psychological judgments are based.

Sensory quality A distinguishing attribute of a stimulus (e.g., sound frequency as the determinant of pitch).

Sensory receptors Convert environmental energy into nervous-system signals that the brain can understand and interpret.

Sensory register Component of the memory system where information is received and held for very short periods of time.

Sensory store The memory store which holds incoming sensory information for a brief period of time. The two most widely investigated are iconic memory (vision) and echoic memory (hearing).

Sensory-motor intelligence In Piaget's theory, intelligence during the first two years of life, consisting mainly of sensations and motor impulses, with little in the way of internalized representations.

Sentence analyzing machinery (SAM) The sequence of strategies that listeners use to comprehend sentences.

Sentencing disparity Inconsistency of sentences for the same offence from one judge to another.

Separation anxiety A set of fearful responses, such as crying, arousal, and clinging to the caregiver, that infants exhibit when the caregiver attempts to leave the infant.

Separation anxiety test A photographic test devised to measure attachment quality in older children and adolescents and differentiate between those who place value on intimate relationships with others, and those who act in ways likely to elicit rejection.

Seplophobia Fear of decaying matter.

Serial ordering The ability to group according to logical progression.

Serial position effect A characteristic of memory retrieval in which the recall of beginning and end items on a list is often better than recall of items appearing in the middle.

Serial processes Two or more mental processes that are carried out in order, one after the other.

Serial reproduction An experimental technique in which a drawing is presented to one research participant, who reproduces it from memory for a second participant, whose reproduction is shown to a third, and so on. Each participant's memory distortions become part of the stimulus for the next participant down the line, such that reconstructive alterations accumulate.

Serial search The successive comparison of a target stimulus to different items in memory.

Seriation Arranging objects in sequential order according to one aspect, such as size, weight, or volume.

Serotonin A neurotransmitter involved in mood, sleep, appetite, and impulsive and aggressive behaviour. Too little has been associated with depression and some anxiety disorders (e.g., obsessive-compulsive disorder). Many antidepressants attempt to reduce the amount of serotonin that is taken back (reuptake) into the sending neuron (e.g., Serotonin Reuptake Inhibitors [SRI]).

Serum cholesterol A fatlike chemical found in the blood. One form (LDL) promotes the formation of atherosclerotic plaques. Another form (HDL) may protect against coronary heart disease.

Servomechanisms Devices that use negative feedback to maintain a stable state of affairs, such as a heating system that maintains a home at a constant temperature.

Sesquipedalophobia Fear of long words.

Set A temporary readiness to perceive or react to a stimulus in a particular way.

Set point The optimum value of the system variable in a regulatory mechanism. The set point for human body temperature, recorded orally, is approximately 98.6°F.

Setpoint hypothesis The hypothesis that different persons have different setpoints for weight, such that attempts to alter one's weight are physiologically countered and, thus, will ultimately prove ineffectual.

Settling down A stage in the culminating phase of early adulthood, ranging from ages 33 to 40.

Sex A biological distinction between males and females.

Sex chromosomes Chromosomes that contain the genes that code for the development of male or female characteristics.

Sex differences Biologically based characteristics that distinguish males from females.

Sex perversions Unnatural forms of sex like homosexuality, fetishism, etc.

Sex related aggression Aggressive behaviour that is elicited by the same stimuli that elicits sexual behaviour. Any person who can evoke sexual desire can equally evoke aggression via jealousy, etc.

Sex role The quality of being male or female, based on anatomy.

Sexism 1. An individual's prejudicial attitudes and discriminatory behaviour toward people of a given sex, or.

2. Institutional practices (even if not motivated by prejudice) that subordinate people of a given sex.

Sexophobia Fear of the opposite sex. (Heterophobia).

Sex-role stereotypes The belief originally found amongst children, including those not yet attending school, about what are most appropriate characteristics for each sex.

Sexual arousal The motivational state of excitement and tension brought about by physiological and cognitive reactions to erotic stimuli.

Sexual dimorphism A characteristic of species, such as humans, where there are two distinguishable forms of male and female, in which males compete over females and are larger than the female or endowed with more dangerous weaponry.

Sexual harassment Unwelcome physical or verbal sexual overtures that create an intimidating, hostile, or offensive social environment.

Sexual identity The sense of being male or female and all that goes with it, including issues of gender identity, gender role, and sexual orientation.

Sexual latency Inactive sexual interest.

Sexual orientation A feeling of attractedness or arousal associated with a particular gender. Sexual behaviour can be a result of this but does not necessarily define a person's orientation.

Sexual sadism An offender who obtains sexual gratification from the victims response to physical/psychological torture.

Sexual script A set of implicit rules that allow a person to judge what sexual behaviour is appropriate in a given situation.

Sexual selection Process in which females choose their mates based on certain characteristics that will then be passed on to their male offspring.

Sexuality Obtaining physical pleasure. A much wider notion than reproduction and one that extends over the entire lifespan. It was apparently Wilhelm Fleiss, an intelligent crank who became Freud's closest friend, who got him thinking about childhood sexuality.

Shading A monocular cue of depth perception; determines whether portions of the surface of an object are perceived as concave or convex.

Shadowing The procedure, often used in dichotic presentations, in which a participant is asked to repeat aloud, word for word, only what she hears through one earphone.

Shallow processing The encoding of a stimulus using its superficial characteristics, such as the way a word sounds or the typeface in which it is printed.

Shamanism A spiritual tradition that involves both healing and gaining contact with the spirit world.

Shape constancy The tendency to perceive objects as retaining their shapes despite changes in our angle of regard that produce changes in the image projected on the retina.

Shaping An instrumental learning procedure in which an animal (or human) learns a rather difficult response through the reinforcement of successive approximations to that response.

Shaping by successive approximations A behavioural method that reinforces responses that successively approximate and ultimately match the desired response.

Shared definition of the situation In the dyad an understanding arrived at by the two individuals involved as to what the nature of their encounter is.

Shared reference The ability of a baby, by the end of its first year, to look at something another person is pointing at.

Shock imparting Giving an electric shock.

Short essay item Test item that includes a question for the student to answer, which may range from a sentence or two to a page of, say, 100 to 150 words.

Short-term memory (STM) Memory processes associated with preservation of recent experiences and with retrieval of information from long-term memory; short-term memory is of limited capacity and stores information for only a short length of time without rehearsal.

Short-term store (STS) The limited capacity storage system supporting conscious mental activity, previously known as primary memory.

Shyness An individual's discomfort and/or inhibition in interpersonal situations that interferes with pursuing interpersonal or professional goals.

Sibling rivalry Based in what Freud, who'd been a momma's boy, saw as a result of childrens' overriding egotism.

Siderodromophobia Fear of trains, railroads or train travel.

Siderophobia Fear of stars.

Sign systems Symbols that cultures create to help people think, communicate, and solve problems.

Signal detection theory (SDT) A systematic approach to the problem of response bias that allows an experimenter to identify and separate the roles of sensory stimuli and the individual's criterion level in producing the final response.

Signature The repetitive ritualistic behaviour of a serial offender. This is typically apparent at every crime scene and has little-to-nothing to do with the perpetration of the crime.

Significant difference A difference between experimental groups or conditions that would have occurred by chance less than an accepted criterion; in psychology, the criterion most often used is a probability of less than 5 times out of 100, or $p < .05$.

Signs In psychopathology, what the diagnostician observes about a patient's physical or mental condition (e.g., tremor, inattentiveness).

Similarity In perception, a principle by which we tend to group like figures, especially by colour and orientation.

Simple affirmative active declaratives (SADD) A term applied to single, simple expressions that are affirmative and declarative, in that they are making positive declarations about something being true and are presented in an active voice.

Simple cell A type of cell in the visual cortex that is orientationally selective and has a relatively small receptive field divided into discrete on and off sub-regions.

Simple phobia An excessive and irrational fear of specific things, such as snakes, darkness, or heights.

Simple random sampling Participants are selected on a completely random basis, such as a random number table. Every member of a population is given an equal chance of being included in a sample.

Simple reaction time A measurement of the speed with which a research participant can respond to a stimulus.

Simple reflex A sensorimotor reflex free of modulation by higher-level influences; approximated by Sir Charles Sherrington, who used spinal animals.

Simplicity The tendency to perceive forms as simple, symmetrical figures rather than as irregular ones.

Simulation An attempt to express an emotion that one does not actually feel.

Simulation heuristic A mental shortcut used in making decisions and evaluating outcomes that employs the imaginary replay of events or situations.

Simulation software Computer programs that model real-life phenomena to promote problem solving and motivate interest in the areas concerned.

Simultaneous colour contrast The effect produced by the fact that any region in the visual field tends to induce its complementary colour in adjoining areas.

For example, a gray patch will tend to look bluish if surrounded by yellow and yellowish if surrounded by blue.

Simultaneous pairing A classical conditioning procedure in which the conditioned stimulus (CS) and the unconditioned stimulus (US) are presented at the same time.

Sine waves Waves that correspond to the plot of the trigonometric sine function.

Single Case Experiment Study of a treatment's effect on one person or one group by contrasting behaviour before, during, and after the treatment is applied.

Single-act technique In the study of personality, the use of a single behaviour to act as a measure for a generalization of all similar situations.

Single-blind study An experiment in which the experimenter but not the subject knows the value of the independent variable.

Single-case experiment A study in which the investigator manipulates the values of some independent variable, just as she would in an experiment with many participants, and then assesses the effects of this variable by recording the single participant's responses.

Single-subject research An experiment or correlational study concerning the behaviour of individual subjects rather than comparisons of the average performance of groups of subjects.

Sinistrophobia Fear of things to the left or left-handed.

Sinophobia.

Sinusoid Alternative word for harmonic or pure tone.

Sitophobia or Sitiophobia.

Situation Modelling the basis for everyday reasoning which is thinking about how things are and how things might be in the world; these can lead to errors in everyday reasoning by being biased or incomplete, or both.

Situational attribute An attribute explained or interpreted as being caused by external influences.

Situational attribution Attribution to factors external to an actor, such as the task, other people, or luck.

Situational demands Events, conditions, or expectations associated with a particular setting.

Situational factors Environmental stimuli that affect a person's behaviour.

Situational variables External influences on behaviour.

Situationism The view that human behaviour is largely determined by the characteristics of the situation rather than personal predispositions.

Sixteen Personality Factor Questionnaire (16PF) A test that assesses sixteen basic dimensions of personality.

Size and shape constancy The consistent size and shape of objects.

Size constancy The tendency to perceive objects as retaining their size, despite the increase or decrease in the size of the image projected on the retina caused by moving closer to or farther from the objects.

Skeletal musculature The muscles (sometimes called *striated muscles*) that are controlled by parts of the somatic nervous system and that move the face, trunk, and limbs.

Skew The degree to which a curve or distribution of scores has extreme scores atypical of the majority of scores.

Skinner box An apparatus developed by B. F. Skinner for observing animal behaviour in experiments in operant conditioning.

Skinner, B. F. Considered the father of behavioural therapy. He once stated that with the ability to control a child's environment, he could raise a child to become anything he wanted.

Sleep apnea A condition in which a person stops breathing many times during a night's sleep.

Sleep deprivation An overall lack of the necessary amount of sleep. A person can be deprived of sleep by their own body and mind, insomnia, or actively deprived by another individual. Sleep deprivation is sometimes used as an instrument of torture, but in recent years it has been shown to be an effective treatment of depression and other mental illnesses.

Sleep paralysis The phenomenon of waking up unable to move for several seconds, due to the persistence of the loss of muscle tone that occurs during REM sleep. While sleep paralysis is sometimes frightening, it is harmless.

Sleep spindles Short bursts of brain waves that occur during stage 2 sleep.

Sleeper effect A delayed impact of a message. Occurs when an initially discounted message becomes effective, as we remember the message but forget the reason for discounting it.

Slow-wave sleep Type of sleep characterized by slow, rolling eye movements, an EEG indicative of low cortical arousal, slowed heart rate and respiration, and recall of "boring," mostly verbal dreams.

Small group discussion A discussion among four to six students in a group working independently of a teacher.

Small muscle development Movements of the fine muscles of the hand.

Smooth muscles The nonstriated muscles controlled by the autonomic nervous system that constrict the blood vessels to help regulate blood pressure and that line many internal organs such as those that produce peristalsis in the digestive tract.

Snowball sampling Find one individual to study who can then connect you to others like him/her; used for hard-to-locate participants.

Social anxiety The unpleasant emotion people experience due to their concern with interpersonal evaluation.

Social categorization The process by which people organize the social environment by categorizing themselves and others into groups.

Social clocks Social and cultural norms that indicate the typical ages at which people experience particular life events, behaviours, and issues.

Social cognition Experiential knowledge and understanding of society and the rules of social behaviour.

Social comparison Evaluating one's abilities and opinions by comparing oneself to others.

Social comparison The process of comparing one's self to others to gather information and to evaluate and judge one's abilities.

Social comparison theory In self-concept formation, the questions of when we make comparisons, what we compare, and why, as well as with whom comparisons are made; this phenomenon is also applied by the individual in social group processes.

Social comparison theory The theory that proposes that we evaluate our thoughts and actions by comparing them to those of others.

Social constructionism A perspective in the social sciences that states that individuals creatively shape reality through social interaction.

Social contract An explanation, used in discussion of the selection task, that specifies that in social exchanges, if one takes a benefit one is expected to pay a cost in return.

Social deprivation The absence of attachment.

Social desirability In the attribution of a disposition the likelihood of a particular action being performed or approved of by most other people in that situation.

Social desirability bias The tendency of some people to describe themselves in socially approved ways.

Social development The ways in which individuals' social interactions and expectations change across the life span.

Social dilemma A problem similar in structure to the prisoner's dilemma but expanded to include any number of individuals, each of whom has to decide whether to cooperate or act individually.

Social exchange theory A theory that asserts that each partner in a social relationship gives something to the other and expects to get something in return.

Social facilitation 1. Original meaning-the tendency of people to perform simple or well-learned tasks better when others are present.

2. Current meaning-strengthening of dominant (prevalent, likely) responses owing to the presence of others.

Social identities Aspects of a person's self-concept based upon his or her group memberships.

Social identity theory In social psychology, the method of examining the group in the context of the individual, where membership of that group is perceived as offering identity to its members, particularly in the context of other, possibly competing, groups.

Social impact theory The theory that the amount of social influence others have depends on their number, strength, and immediacy to those they are trying to influence.

Social inferences Assumptions about the nature of social relationships, processes, and others' feelings.

Social influence A general term for interactions in which an individual's thinking or behaviour is affected by the actions of several others.

Social intelligence A theory of personality that refers to the expertise people bring to their experience of life tasks.

Social interaction The communication and relationships between people that influence the shaping of the self.

Social interaction theory Theory emphasizing the right kind of social interactions for personality development.

Social intimacy Having the same friends and enjoying the same types of recreation.

Social learning theory A theoretical approach to socialization and personality that is midway between radical behaviourism and cognitive approaches to learning. It stresses learning by observing others who serve as models and who show the child whether a response he already knows should or should not be performed.

Social learning view of aggression View that aggression is learned through direct experiences and observations of others behaviours.

Social learning view of prejudice View that prejudice is learned through direct experience, consistent with the manner other attitudes are learned.

Social loafing An example of the diffusion of social impact in which people working collectively on a task generate less total effort than they would had they worked alone.

Social neuroscience The study of the relationship between neural and social processes.

Social norm A general rule of conduct reflecting standards of social approval and disapproval.

Social penetration theory A theory that describes the development of close relationships in terms of increasing self-disclosure.

Social perception A general term for the processes by which people come to understand one another.

Social phobia A disorder characterized by intense anxiety when exposed to certain kinds of social or performance situations.

Social physique anxiety Anxiety about others observing and evaluating one's physique.

Social power The force available to the influencer to motivate attitude or behaviour change.

Social psychology The branch of psychology that studies our social nature or how the actual, imagined, or implied presence of others influences our thoughts, feelings, and behaviours.

Social reality testing The normal process of validating one's beliefs against others. This process is often disrupted in schizophrenia.

Social representations Socially shared beliefs, and widely held ideas and values, including our assumptions and cultural ideologies. Our social representations help us make sense of our world.

Social responsibility norm A societal rule that tells people they should help others who need help even if doing so is costly.

Social role A socially defined pattern of behaviour that is expected of a person who is functioning in a given setting or group.

Social role theory The theory that virtually all of the documented behavioural differences between males and females can be accounted for in terms of cultural stereotypes about gender and the resulting social roles that are taught to the young.

Social schemas Mental models that represent and categorize social events and people.

Social skills Skills or behaviours deemed desirable or necessary to effectively interact with society.

Social skills training A behavioural training program designed to improve interpersonal skills through observation, modeling, role-playing, and behavioural rehearsal.

Social support Resources, including material aid, socioemotional support, and informational aid, provided by others to help a person cope with stress.

Social trap A situation in which one harms oneself and others by acting in one's self-interest.

Socialization The lifelong process whereby an individual's behavioural patterns, values, standards, skills, attitudes, and motives are shaped to conform to those regarded as desirable in a particular society.

Socializing agents Those influences that teach and reinforce society's rules and norms.

Sociobiology A research field that focuses on evolutionary explanations for the social behaviour and social systems of humans and other animal species.

Sociocultural approach Within social psychology and personality psychology, the view that many psychological phenomena, some of which have been presumed to be universal, result from or are affected substantially by cultural norms.

Sociodramatic play A stage of elaborate role-play amongst children over the age of 3 or 4, where pretend games are developed around various characters, such as mothers, fathers and babies, which are often influenced by books, tv programmes and films.

Socioecology The study of how environment influences society in the process of natural selection and which perceives social systems as being flexible and subject to change over evolutionary time-scales.

Socioeconomic status (SES) A measure of prestige within a social group most often based on income and education.

Sociogram A diagrammatic representation of the observation of friendships amongst groups of children.

Sociology Sociology is the study of the social lives of humans, groups, and societies, sometimes defined as the study of social interactions. It is a relatively new academic discipline that evolved in the early 19th century.

Sociometric status-types The categorization of children as 'popular', 'controversial', 'rejected', 'neglected' or 'average' during sociometric studies, according to whether they are high or low on positive and negative nominations.

Sociometry The study of friendships in groups, particularly of children.

Solitary play Play that occurs alone.

Soma The cell body of a neuron, containing the nucleus and cytoplasm of the cell.

Somatic nervous system A division of the peripheral nervous system primarily concerned with the control of the skeletal musculature and the transmission of information from the sense organs.

Somatic therapies A collective term for any treatment of mental disorders by some organic means, including medication, surgery, and electroconvulsive therapy.

Somatization disorder A disorder characterized by a wide variety of physical symptoms, such as pain and gastrointestinal, sexual, and pseudoneurological problems. The disorder begins before age thirty and continues for many years. It is also called hysteria or Briquet's syndrome.

Somatoform disorders The generic term for mental disorders in which bodily symptoms predominate despite the absence of any known physical cause; included are conversion disorder, hypochondriasis, somatization disorder, and somatoform pain disorder.

Somatoform pain disorder A mental disorder in which the sufferer describes chronic pain for which there is no discernible physical basis.

Somatogenic hypothesis The hypothesis that mental disorders result from organic (bodily) causes.

Somatogenic mental disorders Mental disorders that have known organic causes. This is the case for some disorders (e.g., general paresis) but almost surely not for all (e.g., phobias).

Somatomotor system The system of sensory messages that passes from the cerebral cortex to the muscles in the periphery.

Somatosense Bodily sensations; sensitivity to such stimuli as touch, pain, and temperature.

Somatosensory cortex The region of the parietal lobes that processes sensory input from various body areas.

Somatosensory system The system of receptors in the body that communicates the messages relating to sensation and touch.

Soul James Strachey has almost everywhere translated Seele—soul—into "mind" in his Revised Edition of Freud's works and its operations into the medical-sounding "mental apparatus." For Freud, soul was the overarching reality of all psychic functioning, conscious and unconscious. It had no sense for him of something supernatural, but it was his preferred term when describing the psyche as a whole as well as its very essence. If archetypal psychologist James Hillman has placed soul on an imaginal basis, Freud was the first to bring it into the realm of psychological discussion.

Sound waves Changes in pressure generated by vibrating molecules.

Source amnesia Inaccurate recall of the origin of information in memory. It is also called source misattribution or source monitoring error.

Source confusion A type of memory error in which information acquired in one context is remembered as having been encountered in another (e.g., a person's recalling that she had chocolate cake on her last birthday when she actually had it two birthdays ago).

Source memory The knowledge of the event from which a certain memory derived.

Spatial geometry In the ontogeny of environmental cognition as described by piaget, the child's concept of space constructed on topological principles.

Spatial summation The process whereby two or more stimuli that are individually below threshold will elicit a reflex if they occur simultaneously at different points on the body.

Spatial thinking The mental computations engaged in when we must locate objects and discern the spatial relationships among them.

Spearman's Rho A correlational technique used primarily for rank ordered data (ordinal scale).

Spearman's theory of general intelligence (*g*) Spearman's account, based on factor analytic studies of intelligence-test performance, which proposes one underlying factor, general intelligence (*g*), which is tapped by all subtests, and a large number of specific skills (*s*'s) that are tapped by each subtest.

Special education Programs that address the needs of students with mental, emotional, or physical disabilities.

Species-specific behaviour Behaviour patterns characteristic of a species that are typically built-in and emerge without any relevant prior experience.

Species-specific defence reaction Reactions to threat that are largely innate and found in all members of a species, e.g., flight in birds, "playing dead" in opossums.

Species-typical behaviour A behaviour seen in all or most members of a species, such as nest building, special food-getting behaviours, or reproductive behaviours.

Specific language impairment An inherited syndrome that impairs language learning and causes lifelong deficits in sentence comprehension and production, despite otherwise normal cognitive functioning.

Specific phobia A disabling, irrational fear of certain objects or events such as snakes, heights, or enclosed places.

Specificity theory An approach to sensory experience which asserts that different sensory qualities are signaled by different neurons. These neurons are somehow labeled with their quality, so that whenever they fire, the nervous system interprets their activation as that particular sensory quality.

Spectral reflectance The proportion of each incident stimulus wavelength that a surface reflects. Spectral sensitivity the pattern of response of a neurone to different stimulus wavelengths.

Spectral sensitivity The eye's responsiveness to each of the separate wavelengths that constitute a light stimulus.

Spectral sensitivity curve A graphical representation of the eye's spectral sensitivity.

Speech acts In semantic theory, an action that confirms the intention of the party involved in it to do the thing uttered.

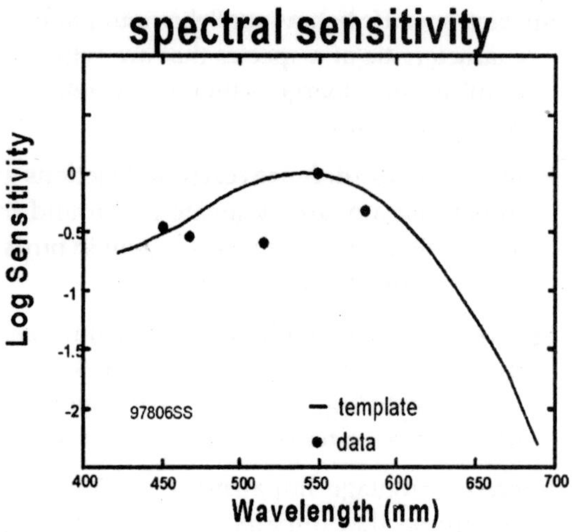

Fig. Speech acts

Speech disorders Articulation problems occurring most frequently among children in the early elementary school grades.

Speech plans Coordinated patterns of speech movements, constructed in Broca's area, that are relayed to the primary motor projection area for decoding into discrete muscular movements.

Sperm (spermatozoom) A sex cell manufactured in the testes and contributed by the male as part of sexual reproduction.

Spinal animal An animal whose spinal cord has been severed in the neck region in order to cut all connections between the body (from the neck down) and the brain, leaving the spinal reflexes free of higher influences.

Spinal cord A long, thin collection of nerve cells attached to the base of the brain and running the length of the spinal column.

Spinal nerve A bundle of nerve fibres attached to the spinal cord; conveys sensory information from the body and carries messages to muscles and glands.

Spinal reflexes Automatic behaviours that require no input from the brain.

Spirituality The power of abstract thought. For this reason Freud saw the commandment against idolatry as an elevation of culture: our object of central concern, God, has become invisible and therefore abstract.

Split brain A condition in which the corpus callosum and some other fibres are cut so that the two cerebral hemispheres are functionally isolated.

Split half reliability The correlation coefficient determined by comparing first half of the measurement to the second half. Measure of the internal consistency of a test or measuring device.

Split-brain operation A surgical procedure that severs the corpus callosum, thus abolishing the direct connections between the cortex of the two cerebral hemispheres.

Spontaneous recovery The tendency for previously learned information to resurface rapidly after a period of extinction. Information that is spontaneously recovered is thought to lay dormant but not forgotten (e.g., riding a bicycle after a long period of not riding).

Spontaneous-remission effect The improvement of some mental patients and clients in psychotherapy without any professional intervention; a baseline criterion against which the effectiveness of therapies must be assessed.

Spreading activation model A memory model that assumes that elements in a semantic memory network are more heavily activated the closer they are to each other within the network.

Stabilized image technique A procedure that projects a stationary image on the retina even though the eye is moving.

Stable attribution An inference that an event or behaviour is due to stable, unchanging factors.

Stable identity The concept that one's self remains consistent even when circumstances change.

Stage A period of development that occurs at about the same time for each person. Developmental and Personality theories are often made up of a series of stages.

Stage theories of development Theories that suggest that people go through a series of discrete stages, each of which is characterized by at least one task that an individual must accomplish before progressing to the next stage.

Stage theory of memory An approach to memory that proposes several memory stores. One is short-term memory, which holds a small amount of information for fairly short intervals; another is long-term memory, which can hold vast amounts of information for extended periods. According to the theory, information can only be transferred to long-term memory if it has first been in short-term memory.

Staging The alteration of a crime scene in order to redirect the investigation in a way away from offender (or at least what the offender thinks logically will do this).

Stagnation The opposite of generativity, this is one of the two polarities of the stage of maturity, as described by erik erikson, when the adult is unable to give to others in an unselfish way and becomes increasingly preoccupied with his/her own concerns.

Standard deviation A measure of spread within a distribution (the square root of the variance). The most popular and most reliable measure of variability but the more skewed a distribution, the more error there will be in the standard deviation because of its reliance on the mean.

Standard error of measurement A statistical procedure used to determine the amount of error of any measurement device.

Standard error of the mean A measure of the variability of the mean whose value depends both on the standard deviation (SD) of the distribution and the number of cases in the sample (N); if SE is the standard error, then SE = SD/.

Standard score A score derived by transforming the data based on the standard deviation. Standard scores can then be compared to one another on face value.

Standardization A set of uniform procedures for treating each participant in a test, interview, or experiment or for recording data.

Standardization group The group against which an individual's test score is evaluated.

Standardization sample A large group of people that is representative of the entire population of potential test takers.

Standardized tests Tests that are usually commercially prepared for nationwide use to provide accurate and meaningful information on student's level of performance relative to others at their age or grade levels.

Stanford Binet Intelligence Scale A popular IQ test.

Stanford-Binet test A revision of the Binet-Simon intelligence test, which provides a scale of ninety items that can be used to test all people, from individuals with learning difficulties to superior adults; the test also provides examples of acceptable and unacceptable answers to test items so that scoring can be carried out with precision.

Stanine A standard score that literally means Standard Nine, stanines have a mean of five and a standard deviation of approximately two. Stanines 2 through 8 are exactly 1/2 standard deviations and stanines one and nine or open ended.

Stanine scores A type of standardized score ranging from 1 to 9, having a mean of 5 and a standard deviation of 2.

State A set of circumstances or attributes characterizing a person or thing at a given time; way or form of being; condition.

State dependent memory The theory that information learned in a particular state of mind (e.g., depressed, happy, somber) is more easily recalled when in that same state of mind.

State specific learning(& State dependent memory) The theory that what is learned in one state mind (example in a certain place, under the influence of a certain drug) is better recalled in a similar state (in the same place, taking the same drug).

State-action tree In problem solving, the complete set of choices available to a problem solver, set out in a diagrammatic form, showing the possible states of the world, represented together with the actions that lead from one state to another.

State-dependent memory The tendency to recall information better when our mental or emotional state at retrieval matches that during encoding.

Statistic An observed characteristic of a sample (e.g., 20% improvement rate, range of IQ's).

Statistical reasoning A form of reasoning applied to arguments that depend on statistics or probability, and which is often important in deciding how to conduct our everyday lives.

Statistical regression The tendency to regress towards the mean score of a group. Those who have extreme scores will most likely tend to be closer to the mean upon retesting. A threat to internal validity.

Statistical reliability The degree to which an observed difference in sample means reflects a real difference in population means and is not attributable to chance.

Statistical significance The likelihood that an observed relation or difference between two variables is not due to chance factors.

Statistics Facts or data of a numerical kind, assembled, classified, and tabulated so as to present significant information about a given subject.

Stem A question or a partial statement in a test item that is completed by one of several choices.

Stem and leaf display A multiple column table depicting the individual digits of the scores. A score of 95 would have a stem of 9 and a leaf of 5, a score of 62 would have a stem of 6 and a leaf of 2. If a particular stem has more than one leaf, such as the scores 54, 58, and 51, the stem of 5 has three leaves, in this case 458.

Stereopsis A form of depth perception based on retinal disparity.

Stereotaxic apparatus A device used to insert an electrode into a particular part of the brain for the purpose of recording electrical activity, stimulating the brain electrically, or producing localized damage.

Stereotype A belief about the personal attributes of a group of people. Stereotypes are sometimes overgeneralized, inaccurate, and resistant to new information.

Stereotype threat A disruptive concern, when facing a negative stereotype, that one will be evaluated based on a negative stereotype. Unlike self-fulfilling prophecies that hammer one's reputation into one's self-concept, stereotype threat situations have immediate effects.

Stereotype vulnerability A disturbing awareness among members of a negatively stereotyped group that anything one does, or anything about oneself that fits the stereotype, may confirm it as a self-characterization.

Stereotyping The imposition, as a result of irrational prejudices, of a set of simplified characteristics upon a racial, social or other person or group, despite evidence to the contrary.

Stigma An attribute that serves to discredit a person in the eyes of others.

Stillbirth The birth of a dead foetus after 20 weeks.

Stimulant An influence (typically, a drug) that has activating or excitatory effects on brain or bodily functions (e.g., amphetamines, Ritalin, cocaine).

Stimulus In psychology, anything effectively impinging upon any of the sensory apparatuses of a living organism, including physical phenomena both internal and external to the body.

Stimulus discrimination In classical conditioning, the tendency *not* to have a conditioned response to a new stimulus that's similar to the original conditioned stimulus. In operant conditioning, it's the tendency for a response to happen only when a particular stimulus is present.

Stimulus equivalence A type of learning in which stimuli become equivalent even though the organism has never observed a relation between them; may be involved in learning how to read and manipulate symbols.

Stimulus generalization In classical conditioning, the tendency to respond to stimuli other than the original conditioned stimulus (CS). The greater the similarity between the CS and the new stimulus (CS+), the greater generalization will be. An analogous phenomenon in instrumental conditioning is a response to stimuli other than the original discriminative stimulus.

Stimulus segmentation The organization of a stimulus into parts that belong together and parts that don't and which is also known as grouping or parsing.

Stimulus-driven capture A determinant of why people select some parts of sensory input for further processing; occurs when features of stimuli—objects in the environment—automatically capture attention, independent of the local goals of a perceiver.

Storage The process of saving information in long term memory.

Storage capacity The amount of information that can be retained in memory.

Strange situation The procedure used with infants aged 1 to 3 years to measure the relationship between an infant and a particular care giver, which assesses the infant's response to being mildly stressed by being taken from them and left with a stranger.

Stranger Anxiety distress in the presence of unfamiliar people.

Stranger anxiety The wariness and fearful responses, such as crying and clinging to their caregivers, that infants exhibit in the presence of strangers.

Strategic retrieval A deliberate effort to recall information by supplying one's own retrieval cues (e.g., "Let's *see*, the last time I remember seeing my wallet was . . .").

Strategic self-presentation Conscious and deliberate efforts to shape other people's impressions in order to gain power, influence, sympathy or approval.

Stratified random sampling The researcher samples equally from each group of distinctly different individuals (strata) in a population.

Stratified sampling An experimental procedure in which each subgroup of the population is sampled in proportion to its size.

Strepsirhines One of the two great branches of the primate order, comprised of lemurs and lorises.

Stress In psychology, a physical or psychological stimulus such as very high heat, public criticism, or another noxious agent or experience which produces psychological strain.

Stress inoculation training A method of teaching people to cope with anticipated problems. It involves three steps: education, rehearsal, and application.

Stress moderator variables Variables that change the impact of a stressor on a given type of stress reaction.

Stressor An internal or external event or stimulus that induces stress.

Striving for superiority The motivation to seek superiority. Adler argued that striving for superiority is born from our need to compensate for our inferiorities.

Stroke A cerebrovascular accident; damage to the brain caused by a blood clot in a cerebral artery or rupture of a cerebral blood vessel.

Stroop effect A marked decrease in the speed of naming the colours in which various colour names (such as green, red, etc.) Are printed when the colours and the names are different. An important example of automatization.

Structural encoding A way of encoding verbal information that emphasizes how words look.

Structural family therapy A form of family therapy in which the maladaptive relationships among family members is inferred from their behaviour and attempts are made to restructure these behaviours into more adaptive ones.

Structuralism The view that all human mental experience can be understood as the combination of simple events or elements, and that the underlying structure of the human mind can be revealed by analyzing all basic elements of sensation. The study of the how and why of experience, associated with Wilhem Wundt.

Structured interview A set of face-to-face questions posed to a subject in social research.

Structured personality test A personality test (e.g., the MMPI or CPI) that asks specific questions and requires specific answers.

Student Teams-Achievement Divisions(STAD) A cooperative learning method for mixed-ability groupings involving team recognition and group responsibility for individual learning.

Students at risk Students who are subject to school failure because of characteristics of the student or inadequate responses to their needs by school, family, or community.

Stygiophobia or Stigiophobia Fear of hell.

Subconscious awareness The mental processes involving information not currently in consciousness but retrievable by special recall procedures, such as hypnosis.

Subcortical The lower areas of the brain, which are responsible for basic life functions.

Subcortical structures Usually the forebrain structures, such as those comprising the limbic system and extrapyramidal motor system, that lie beneath the cortex.

Subculture A social group exhibiting a lifestyle sufficiently different to distinguish itself from others within the larger culture.

Subject An individual person or animal that a researcher studies.

Subject bias Bias that results from the subject's expectations or the subject's changing of his or her behaviour.

Subject matching A method of reducing bias in a sample of subjects by matching specific criteria of the sample to the true characteristics of the population. (Example: If the population is 60% female then 60% of the subjects in the sample should also be female).

Subject variable A variable that characterizes pre-existing differences among the participants in a study.

Subjective contours Perceived contours that do not exist physically. We tend to complete figures that have gaps in them by perceiving a contour as continuing along its original path.

Subjective norm In the theory of reasoned action, a person's perceptions of social approval for their performing a particular behaviour and which will serve as a predictor of the action taken.

Subjective reality The perception of reality made by an individual that may be different from the perception made by another person.

Subjective utility The process of making a decision by estimating the personal value of a decision's outcome.

Subjective values The outcomes an individual prefers; in Walter Mischel's view, a determinant of behaviour and an important individual difference.

Subjective well-being One's happiness, or life satisfaction, as measured by self-report.

Sublimation When a sexual drive is directed into a different aim, like daily work, creativity, piety, etc. (Jung didn't like that Freud came up with this concept, so he coïned a term that means pretty much the same thing but called it psychization.) It was Goethe who introduced this term into common German usage; Hegel took it up later, as did Freud.

Subliminal perception The perception of a stimulus, as indicated by a change in behaviour, at an intensity insufficient to produce a conscious sensation.

Subordinate concept A concept that refers to types of items within a basic-level category.

Subroutines In a hierarchical organization, lower-level operations that function semiautonomously but are supervised by higher-level ones.

Substance abuse According to the *DSM*, a maladaptive pattern of drug use that results in repeated negative consequences such as legal, social, work-related, or school-related problems.

Subtractive colour mixture Colour that results from the subtraction of one set of wavelengths from another set, commonly produced when mixing colour pigments or superimposing two coloured filters.

Subtyping Accommodating individuals who deviate from one's stereotype by splitting off a subgroup stereotype (such as "middle class blacks" or "feminist women"). Subtyping protects stereotypes.

Subvocal articulation An unvoiced speech utterance.

Success for All A comprehensive approach to prevention and early intervention for preschool, kindergarten, and grades 1 through 5, with one-to-one tutoring, family support services, and changes in instruction that might be needed to prevent students from falling behind.

Successive approximations The process of shaping a response by rewarding the steps toward the behavioural goal.

Sudden infant death syndrome (SIDS) The sudden death of an apparently healthy infant, of unknown cause but believed to be related to some faulty mechanism in respiration control.

Suggestibility In the study of crowd behaviour, the responsiveness of individuals within a crowd to taking part in often violent group activity, as if in a hypnotic trance.

Suicide The act of killing oneself intentionally.

Sulci The clefts in the cerebrum.

Summarization Brief statements that represent the main idea of the information being read.

Summative evaluation Assessments that follow instruction and evaluate knowledge or skills.

Summative quiz Final test of an objective.

Sunk cost principle The economic rule of thumb that only future costs and benefits, not past commitments, should be considered in making a decision.

Super ego In Freudian psychology, one of the fundamental components of the personality; the super-ego develops in the child from the age of 3 years, emerging out of the ego as the seat of the conscience and controlling instinctive desires as a result of the learning of standards of social behaviour from the parent.

Superior conformity of the self In social comparison processes, the belief that everyone within a particular group has that their position is like the others', only a little better.

Superordinate concept A concept that refers to collections of basic-level concepts.

Superordinate goal A shared goal that can be achieved only through cooperation among individuals or groups.

Superstitious behaviour A behaviour that occurs in response to the regular, noncontingent occurrence of an appetitive stimulus to a motivated organism; appears to cause a certain event, but in reality does not.

Supertraits Hans Eysenck's term for his two distinct categories of personality traits. They include Introversion-Extroversion and Neuroticism. According to Eysenck, each of us fall on a continuum based on the degree of each supertraits.

Supplication Advertising one's weaknesses or one's dependence upon others in order to solicit help or sympathy.

Suppression The defence mechanism where we push unacceptable thoughts out of consciousness and into our unconscious.

Suprachiasmatic nucleus The main biological clock regulating circadian rhythms of sleep in humans.

Surface dyslexia A reading disorder in which people can read words phonetically but have difficulty reading irregularly spelled words by the whole-word method.

Surface reflectance The proportion of incident light that a given surface will reflect, rather than absorb, also known as albedo, and which is the physical correlate of perceived lightness.

Surface structure The grammatical features of a sentence. See also *deep structure*.

Surgency Trait characterized by being outgoing, socially dominant, and high on positive affect; see also extraversion.

Suriphobia Fear of mice.

Surround subregion One of two concentric, circular regions in which the receptive fields of retinal ganglion cells are typically arranged.

Survey A method of getting information about a specific behaviour, experience, or event by means of interviews or questionnaires, using several participants.

Survey research Research that involves interviewing or administering questionnaires to large numbers of people.

Survival advantage The outcome of a characteristic that helps an organism to live long enough to reproduce and pass on its genes.

Survival of the Fittest A key concept in the theory of evolution; the idea that those organisms best adapted to their environment will be more successful than those who are not. Although a concept from biology, survival of the fittest, and natural selection began to be used in other spheres as early as 1900.

Sustained response The type of neural response that continues throughout the presentation of a stimulus.

Syllogism A deductive argument with two premises, each of which must take one of four forms, and which must also take one fo the same four forms as its conclusion.

Syllogistic reasoning In psychology, a form of reasoning based on the used of syllogisms.

Symbol A sound, gesture, or written character that represents an object, action, event, or idea.

Symbolic interaction theory A contemporary sociological theory, inspired by Mead's insights and based on the premise that the self and social reality emerge due to the meaningful communication among people.

Symbolic reasoning The cognitive ability to relate one concept to another that represents it in some way. For example, a young child's ability to reason symbolically can be tested by placing a small doll in a model room, and then asking the child to find the full-size doll in an analogous place in a normal-size room.

Symbolic representation A type of mental representation that does not correspond to the physical characteristics of that which it represents. Thus, the word *mouse* does not resemble the small rodent it represents.

Symbolic thought The ability to represent objects in terms of mental symbols.

Symbolophobia Fear of symbolism.

Symmetrical distribution A distribution of numerical data in which deviations in either direction from the mean are equally frequent.

Symmetrophobia Fear of symmetry.

Sympathetic division The subdivision of the autonomic nervous system that deals with emergency response and the mobilization of energy.

Sympathetic nervous system (SNS) Part of the Autonomic Nervous System responsible for the fight or flight phenomenon and which plays a role (along with the Parasympathetic Nervous System) in maintaining the body's homeostasis.

Sympathetic vibration The principle that similar minds communicate to and influence each other psychically, like when a string in a one musical instrument is sounded another string tuned alike sounds by resonance.

Symptom A partial satisfaction of a repressed wish. Symptoms blend repressive mechanisms with symbolic satisfactions; for that reason Freud considered them compromise formations. Every symptom has some anxiety behind it.

Synapse The juncture of two neurons, consisting of the presynaptic and postsynaptic membranes, and – in nonelectrical synapses – the synaptic gap between them.

Synaptic cleft A fluid-filled gap between the presynaptic and postsynaptic membranes; the terminal button releases transmitter substance into this space.

Synaptic transmission The relaying of information from one neuron to another across the synaptic gap.

Synaptic vesicle A small spherical packet located within each terminal bouton and which contains neurotransmitters.

Syndrome A pattern of signs and symptoms that tend to co-occur.

Syngenesophobia Fear of relatives.

Synomorphy In environmental psychology, the fit between the activities that go on within a behaviour setting and the actual physical arrangements of that setting.

Syntactical rule A grammatical rule of a particular language for combining words to form phrases, clauses, and sentences.

Syntax The system by which words are arranged into meaningful phrases and sentences.

Syphilophobia Fear of syphilis.

System variable The variable controlled by a regulatory mechanism; for example, temperature in a heating system.

Systematic desensitization A behaviour therapy that tries to remove anxiety connected to various stimuli by gradually counterconditioning to them a response incompatible with fear, usually muscular relaxation. The stimuli are usually evoked as mental images according to an anxiety hierarchy, whereby relaxation is conditioned to the less frightening stimuli before the more frightening ones.

T

T lymphocytes Cells that develop in the thymus gland that produce antibodies, which defend the body against fungi, viruses, and multicellular parasites.

T score A standard score that sets the mean to fifty and standard deviation to ten. Used on a number of tests including the MMPI.

T test for dependent means the type of t-test used when each score in a one set of data is paired with scores in another set of data.

T test for independent means the type of t-test used when the groups or samples are not related in some relevant way.

Table of specifications List of instructional objectives and expected levels of understanding that guide test development.

Taboo A sacred-feeling prohibition that applies to an entire group. Its true purpose is to keep incestuous wishes unconscious. The rituals surrounding the taboo are similar to those of obsessive-compulsive neurosis but are collective rather than personal.

Tachistoscope A device that can present visual stimuli for controlled (usually very brief) durations of time.

Tachophobia Fear of speed.

Tacit knowledge Practical "how to" knowledge that is unwittingly accumulated from everyday experience.

Tactile The sense of touch. .

Taeniophobia or teniophobia Fear of tapeworms.

Taoism A religion in China renovated by Lao Tsu in 6th century B.C., emphasizing harmony with nature.

Tape-recorder theory of memory The erroneous view that the brain contains an indelible record of everything one experiences.

Taphephobia taphophobia Fear of being buried alive or of cemeteries.

Tapinophobia Fear of being contagious.

Tardive dyskinesia A serious movement disorder that can occur when a person has been treated with antipsychotic drugs for an extended period.

Tardive dyskinesia A serious side effect of antipsychotic drugs. It is usually a permanent condition, characterized by involuntary movements.

Target cell A cell whose physiological processes are affected by a particular hormone; contains special receptor molecules that respond to the presence of the hormone.

Task analysis Breaking down tasks into fundamental subskills.

Taste bud A small organ on the tongue that contains a group of gustatory receptor cells.

Taste-aversion learning A biological constraint on learning in which an organism learns in one trial to avoid a food whose ingestion is followed by illness.

Taurophobia Fear of bulls.

Taxonomy A classification system.

Taxonomy of educational objectives Bloom's ordering of objectives from simple learning tasks to more complex ones.

T-data A category of data used in personality assessment, relating to objective tests of behaviour.

Teaching objectives Clear statement of what students are intended to learn through instruction.

Technical eclecticism In the practice of psychotherapy, the use of whatever techniques work best for particular persons and problems. This approach contrasts with views that therapy should proceed according to a consistent therapeutic orientation (e.g., psychodynamic, behavioural, etc.).

Technophobia Fear of technology.

Tectorial membrane A membrane located above the basilar membrane; serves as a shelf against which the cilia of the auditory hair cells move.

Telegrammatic speech The way in which the young child under 3 selectively omits grammatical markers such as '-ing', '-ed', and words which carry relatively little information, such as 'the', 'and' and 'is' thus creating an abbreviated style of speech.

Telegraphic speech Speech of 1 to 2-year-olds in which two or more meaningful words are put together to form brief sentences.

Telegraphic stage The last stage in early language development, when a child begins to form simple sentences and maintains a cognitive word order that can be understood as reflecting a native language.

Teleophobia .1 Fear of definite plans. .2 Religious ceremony.

Telepathy Freud believed tentatively in the possibility of telepathy, linking it with the emergence of psychic material from the primary to the secondary process.

Telephonophobia Fear of telephones.

Temperament The varying categories of behaviour in young babies which may persist and contribute to the development of personality in adulthood.

Template A hypothetical pattern that resides in the nervous system and is used to perceive objects or shapes by a process of comparison.

Temporal coding A means of representing information by the nervous system; different features are coded by the pattern of activity of neurons.

Temporal contiguity Co-occurrence of stimuli. A condition Pavlov thought would be favourable for forming associations; actually forward pairing is most favourable.

Temporal lobe The lobe of the cortex lying below the temples in each cerebral hemisphere, which includes the primary auditory projection area, Wernicke's area, and subcortically, the amygdala and hippocampus.

Temporal summation The process whereby a stimulus that is below threshold will elicit a reflex if the stimulus occurs repeatedly.

Tend-and-befriend response A response to stressors that is hypothesized to be typical for females; stressors prompt females to protect their offspring and join social groups to reduce vulnerability.

Teratogens A class of environmental hazards which cause marked abnormalities in the foetus during pre-natal development and which are can be caused by the use of harmful drugs or substances, such as thalidomide and cocaine, cigarette smoking and heavy use of alcohol.

Teratophobia Fear of bearing a deformed child or fear of monsters or deformed people.

Terminal bouton A swelling located at the end of an axon which butts on to the dendrite or cell body of other nerve cells.

Territorial aggression Threat or attack behaviour displayed towards an invasion of ones territory or the submissive-retreat behaviour displayed when confronted while intruding.

Territorial instinct Instinctual need to have a certain amount space around oneself.

Territory A term used by ethologists to describe a region a particular animal stakes out as its own. The territory holder is usually a male, but in some species the territory is held by a mating pair or by a group.

Test The most formal kind of self-report, which is normally highly quantitative in nature with the subject's responses being tightly controlled.

Test bias An undesirable characteristic of tests in which item content discriminates against certain students on the basis of socioeconomic status, race, ethnicity, or gender.

Test profile A graphical indication of an individual's performance on several components of a test. This is often useful for guidance or clinical evaluation because it indicates that person's pattern of abilities or traits.

Test retest reliability The correlation coefficient determined by comparing the scores of the same measuring device administered to the same people on two different occasions.

Testing The physical or menta effects that a first test has on the subject's performance on subsequent identical tests. A threat to internal validity.

Testophobia Fear of taking tests.

Testosterone The male sex hormone, secreted by the testes, that stimulates production of sperm and is also responsible for the development of male secondary sex characteristics.

Test-retest method A way of ascertaining the reliability of a test. It involves administering the same test to the same group of subjects after a certain time lag and assessing the correlation of scores.

Test-retest reliability The ability of a test to produce the same results when given to the same group of people at different times.

Tetanophobia Fear of lockjaw, tetanus.

Tetrachoric correlation A correlational technique used to estimate the Pearson-Product correlation of two continuous variables that have been dichotomized (Example: age is continuous, but when it is split into two groups, such as over 40 and under 40, it becomes dichotomous).

Tetrahydrocannabinol (THC) The hallucinatory chemical that is the principal and most active ingredient in marijuana.

Teutophobia Fear of German or German things.

Textophobia Fear of certain fabrics.

Texture A monocular cue of depth perception; the fineness of detail present in the surface of objects or in the ground or floor of a scene.

Texture elements Surface variations that indicate the texture of an object (e.g., pebbles on a trail or blades of grass on a lawn) and whose spacing can be used to judge an object's size or one's distance from the object.

Texture gradient A distance cue based on changes in surface texture that depend on how far away the observer is.

Thaasophobia Fear of sitting.

Thalamus A structure located at the top of the brain stem in the diencephalon, that is subdivided into several nuclei that act as relays in the transmission of information such as sensation and pain to and from other brain regions.

Thalassophobia Fear of the sea.

Thanatologists Those who examine all aspects of death, including biological, psychological, and social issues.

Thanatology The study of death, especially of the medical, psychological, and social problems associated with dying.

Thanatophobia or Thantophobia Fear of death or dying.

Thanatos The mythic name students of Freud gave to the death drive he postulated in 1920 in Beyond the Pleasure Principle. This drive represents the organic need to return to lifelessness and stasis, the ultimate calm of lifeless non-conflict. Freud traced all aggressive and destructive activity to this notion, which impressed him deeply after the outbreak of one world war and the death of his daughter Sophie. That portion of the drive which is turned outward benefits the organism, which would otherwise direct it against itself. However, Freud did concede that destructiveness also affords the ego satisfaction of its old narcissistic need for omnipotence.

That's not-all technique A two-step compliance technique in which the influencer begins with an inflated request, then decreases its apparent size by offering a discount or bonus.

Theatrophobia Fear of theatres.

Thematic apperception test (TAT) A projective test in which a person is shown a series of ambiguous pictures that involve people. The person is asked to make up a story about what the people are doing or thinking. The person's responses are believed to reflect aspects of his or her personality.

Theologicophobia Fear of theology.

Theophobia Fear of gods or religion.

Theory A coherent collection of principles used in the science of psychology to explain phenomena. A substantial proportion of theories in psychology are related to the mechanisms by which things work.

Theory of ecological optics A theory of perception that emphasizes the richness of stimulus information and views the perceiver as an active explorer of the environment.

Theory of mind Also known as mind reading, the process described by some psychologists, and still the subject of considerable debate, whereby children develop hypotheses about other people's emotions, desires and beliefs.

Theory of natural selection A theory that explains the process of evolution. It states that inherited characteristics that give an organism a reproductive or survival advantage are passed on more often to future generations than other inherited characteristics.

Theory of planned behaviour The theory that people's conscious decisions to engage in specific actions are determined by their attitudes toward the behaviour in question, the relevant subjective norms, and their perceived behavioural control.

Theory of psychological reactance The theory that people believe they possess specific behavioural freedoms, and that they will react against and resent attempts to limit this sense of freedom.

Theory of self efficacy Albert Bandura's concept of an individual's belief that he or she can perform adequately in a particular situation.

Therapeutic alliance In psychotherapy, the structuring of the patient-therapist relationship such that the patient considers the therapist a sympathetic ally when confronting problems.

Therapeutic window The amount of a drug that is required for an effect without toxicity.

Thermophobia Fear of heat.

Thermoreceptors Specialized neurons that respond to the temperature of bodily fluids circulating throughout the brain.

Thermoregulation The process by which organisms maintain a constant body temperature. For ectotherms, it is matter of external behaviour such as seeking sun or shade, while for endotherms it also involves numerous internal adjustments such as sweating.

Theta activity EEG activity of 3.5-7.5 Hz; occurs during the transition between sleep and wakefulness.

Theta waves The type of brain waves present when a person is lightly asleep.

Think-aloud protocol A useful source of psychological data which although similar to an introspection does not allow for speculation about process and which usually collects behaviour in an experimental or highly controlled setting, by asking people to 'say what is in their mind' as they solve problems.

Third-variable problem The major obstacle to discerning causation from correlation, because two variables may be correlated only because of the operation of a third variable. For example, sales of ice cream are correlated with rates of violent crime, but only because both increase during hot weather and decrease during cold weather.

Thog problem A problem devised as part of the selection task, involving four geometric figures with different shapes and colours, where the subject has to work through the consequences of each possibility to see how it classifies the other figures in the group.

Thought disorder A pattern of disorganized, illogical, and irrational thought that often accompanies schizophrenia.

Thought stopping A behaviour modification technique in which critical and independent thinking is discouraged or disallowed.

Threat displays In animals, dramatic patterns of behaviour (e.g., movements, calls, colouration changes) that indicate the intent to fight if challenged further. Threat displays often allow animals to circumvent actual combat, which is far more costly.

Threat gesture A stereotyped gesture that signifies that one animal is likely to attack another member of the species.

Threat-to-self-esteem model A theory stating that if receiving help contains negative self-messages, recipients are likely to feel threatened and respond negatively.

Three-term contingency The relation among discriminative stimuli, behaviour, and the consequences of that behaviour. A motivated organism emits a specific response in presence of a discriminative stimulus because in the past, that response has been reinforced only when the discriminative stimulus is present.

Three-term series problem A kind of deductive reasoning problem, which has two statements about three people or objects that can be ordered according to a relation such as height.

Threshold The point at which a stimulus, or a change in the value of a stimulus, can just be detected.

Tilt after effect The temporary visual distortion of perceived orientation following adaptation to an oriented stimulus.

Timbre A perceptual dimension of sound, determined by the complexity of the sound; for example, as shown by a mathematical analysis of the sound wave.

Timbre perceived sound Quality, corresponding roughly to the physical shape of each cycle of a periodic sound waveform or, equivalently, to the shape of its Fourier spectrum.

Time on task Time spent actively engaged in learning the task at hand.

Time out Removing a student from a situation in which misbehaviour was reinforced.

Time series design A research design where subjects are measured at specific times before and after the treatment has been administered in order to determine the long term effects of the treatment.

Time theory of pitch perception The idea that perceived pitch depends upon the temporal pattern of activity in the auditory nerve.

Timing cue One of the main sources of information in locating the direction of sound sources.

Tip of the tongue phenomenon An occasional problem with retrieval of information that we are sure we know but cannot immediately remember.

Title I Formerly Chapter 1, compensatory programs that were reauthorized as Title 1 of the Improving America's Schools Act (IASA) in 1994.

Tocophobia Fear of pregnancy or childbirth.

Toddlerhood ages 1 to 2.

Token economy An arrangement for operant behaviour modification, usually in institutional settings. Certain responses (e.g., cleaning one's room, talking to others) are reinforced with tokens that can be exchanged for desirable items.

Tolerance A situation that occurs with continued use of a drug in which an individual requires greater dosages to achieve the same effect.

Tomophobia Fear of surgical operations.

Tonitrophobia Fear of thunder.

Tonotopic map The orderly projection of neurones from the cochlea so that nearby cells at the destination are tuned to nearby frequencies.

Top-down processes Processes in form recognition that begin with larger units and then proceed to smaller units (e.g., from phrases to words to letters). This contrasts with bottom-up processes, which start with smaller component parts and gradually build up to the larger units (e.g., from letters to words to phrases). One demonstration of top-down processing is provided by context effects in which knowledge or expectations affect what one sees.

Topicalization A feature of all languages, and which is very common in some dialects, in which some phrase, which appears logically to belong buried deep in a sentence, is moved to the front of the sentence for emphasis.

Topographical Freud's division of psyche into three layers: the preconscious, the conscious, and the unconscious.

Topophobia Fear of certain places or situations, such as stage fright.

Totem A common ancestor of a clan. Symbolized most often by an animal, it symbolizes the clan's guardian spirit or helper. Psychologically, the totem is a stand-in for the father.

Totemism The earliest form of religion, intended to forbid incestuous desires felt by family members toward one another. Freud regarded it as the basis of all other social obligations and norms.

Tourette's syndrome A neurological disorder characterized by tics and involuntary utterances, some of which may involve obscenities and the repetition of others' utterances.

Toxiphobia or Toxophobia or Toxicophobia Fear of poison or of being accidently poisoned.

Trace consolidation hypothesis The hypothesis that newly acquired memory traces undergo a gradual change that makes them more and more resistant to any disturbance.

Tracks Classes or curricula targeted for students of a specified achievement or ability level.

Trait In biology, a trait or character is a genetically inherited feature of an organism. As synonym for trait can be sometimes used the term phenotype; the difference is that this second term can also indicate the total physical appearance or constitution.

Trait theory The view that differences in personality are best characterized in terms of underlying, possibly innate, attributes (traits) that predispose one toward patterns of thinking and behaviour that are essentially consistent over time and across situations.

Trance An altered state of consciousness, a state of mind in which consciousness is fragile and voluntary action is poor or missing; a state resembling deep sleep.

Transactional leader A leader who gains compliance and support from followers primarily through goal setting and the use of rewards.

Transactive memory A collectively shared memory system for encoding, storing, and retrieving information that is greater than any individual memories.

Transcranial magnetic stimulation (TMS) A noninvasive procedure for treating severe depression that involves stimulation of the brain by means of a magnetic coil.

Transduction The process by which a receptor reacts to some physical stimulus (e.g., light or pressure) and creates action potentials in another neuron.

Transection Surgical cutting of a nerve tract or brain region, performed to isolate functionally the regions on either side.

Transfer appropriate processing A theory that proposes that memory is stronger and lasts longer when the conditions of performance are similar to those under which learning occurred.

Transfer of learning The application of knowledge acquired in one situation to new situations.

Transfer of training tests Procedures used to ascertain whether skills learned in one setting generalize to other settings.

Transfer-appropriate processing The perspective that suggests that memory is best when the type of processing carried out at encoding matches the processes carried out at retrieval.

Transference A type of **projection** in which early parental conflicts are reexperienced with the therapist, whose job is to interpret them back to the patient. Freud first saw transference as a hindrance because it distorted the relationship between patient and therapist; later, he argued that a positive transference onto the analyst could help the psychoanalysis progress. He distinguished three kinds: negative, erotic, and sensible.

Transformation Making a series of changes to achieve a specific goal.

Transformational grammar The rules governing a proposition by Noam Chomsky that the most basic sentences, essentially sadds are transformed into more complex sentences by a sequence of one or more transformations.

Transformational leader A leader who changes (transforms) the outlook and behaviour of followers (also referred to as a charismatic leader).

Transient response The type of neural response that occurs only at the onset or offset of a stimulus.

Transitvity A skill learned during the concrete operational stage of cognitive development in which individuals can mentally arrange and compare objects.

Transmitter substance A chemical released by the terminal buttons that causes the postsynaptic neuron to be excited or inhibited.

Transposition The phenomenon whereby visual and auditory patterns (i.e., figures and melodies) remain essentially the same even though the parts of which they are composed change.

Traumatic brain injury (TBI) Direct injury to the brain, such as a tearing of nerve fibres, bruising of the brain tissues against the skull, brain stem trauma, or swelling.

Traumatic neuroses Those caused by situations of unusually high stress. They are the sole possible exception to the rule of neuroses being caused by unresolved Oedipal conflicts.

Traumatophobia Fear of injury.

Travelling wave The dynamic distortion of the basilar membrane in the cochlea produced by auditory stimuli.

Treatment A special program that is the subject of an experiment.

Treatment group Experimental participants who are exposed to nonzero levels of the independent variable.

Tree diagram A branched diagram depicting a hierarchical structure.

Tremophobia Fear of trembling.

Trephining Drilling or cutting holes in the skull. Purportedly once done by demonologically oriented practitioners to allow the escape of evil spirits that were believed to be causing mental disorders.

Trial and error learning Learning that takes place through the application of possible solutions to a problem.

Triangular theory of love A theory proposing that love has three basic components—intimacy, passion, and commitment—which can be combined to produce eight subtypes.

Triarchic theory of intelligence A theory stating that intelligence consists of three factors. Information-processing skills, context, and experience.

Trichinophobia Fear of trichinosis.

Trichopathophobia or Trichophobia Fear of hair. (Chaetophobia, Hypertrichophobia).

Trichromatic theory A theory of colour vision that states that there are three different types of cones in the retina, which are sensitive to light of three different wavelengths. It is also called the *Young-Helmholtz theory*.

Tricyclic antidepressants (tricyclics) An early class of antidepressant medications that includes Tofranil and Pamelor. Because of unfavourable side effects, they have been largely supplanted by selective serotonin reuptake inhibitors.

Tricyclics A class of antidepressant drugs that increase the level of norepinephrine and serotonin.

Trimester A period or term of three calendar months (13 weeks).

Trios Psychologist James Jones's theory that the residual influences and harsh experiences of slavery surface in some African-Americans' conceptions of time, rhythm, improvisation, speech, and spirituality.

Triskaidekaphobia Fear of the number 13.

Tritanopia A form of hereditary anomalous colour vision; caused by a lack of "blue" cones in the retina.

Trophoblast A layer of nutritive ectoderm outside the blastoderm, by which the fertilized ovum is attached to the uterine wall and the developing embryo receives its nourishment.

Tropophobia Fear of moving or making changes.

True experiment Research design that utilizes the most control over subjects and utilizes randomization.

True experimental design A design in which treatments are given and participants are randomly assigned to groups.

True hermaphroditism Rare type of intersexuality in which an individual possesses reproductive tissue of both sexes (e.g., testes and ovaries).

True score the amount of the observed score that truly represents what you are intending to measure.

True-false item One form of multiple-choice test item, most useful when a comparison of two alternatives is called for.

Trust The polar opposite of mistrust, which the young child learns to master as part of the development of ego functioning.

Trypanophobia Fear of injections.

Tuberculophobia Fear of tuberculosis.

Tutorial programs Computer programs that teach lessons by varying their content and pace according to student responses.

Twin studies Studies in which researchers examine trait similarities between identical and fraternal twin pairs to figure out whether that trait might be inherited.

Two tailed The type of test used with a non-directional hypothesis, using +/- 1.96 as the critical value.

Two-factor theory The idea that people's experience of emotion depends on two factors: physiological arousal and the cognitive interpretation of that arousal. When people perceive physiological symptoms of arousal, they look for an environmental explanation of this arousal.

Two-point discrimination threshold The minimum distance between two small points that can be detected as separate stimuli when pressed against a particular region of the skin.

Two-step flow of communication The process by which media influence often occurs through opinion leaders, who in turn influence others.

Two-syndrome hypothesis of schizophrenia The hypothesis that there may actually be two supercategories of schizophrenia, Type I and Type II. According to the hypothesis, Type I is produced by a malfunction of transmitters, especially dopamine, and produces primarily positive symptoms, while Type II is accompanied by reduced frontal lobe activity and sometimes cerebral atrophy, which lead to negative symptoms.

Two way ANOVA An Analysis of Variance used when there are two independent variables.

Two word stage The early stage of language development when a child begins to use phrases to express common functions, such as locating and naming objects, demanding and desiring things, questioning, modifying, and qualifying.

Type A behaviour pattern A complex pattern of behaviours and emotions that includes excessive emphasis on competition, aggression, impatience, and hostility; hostility increases the risk of coronary heart disease.

Type A personality A theory used to describe a person with a significant number of traits focused on urgency, impatience, success, and excessive competition.

Type B behaviour pattern In contrast to the Type A pattern, Type B's are easygoing, less hurried, less competitive, and friendlier; they are less predisposed to coronary heart disease.

Type B personality A theory used to describe person with a significant number of traits focused on relaxation, lack of urgency, and normal or reduced competition.

Type C behaviour pattern A constellation of behaviours that may predict which individuals are more likely to develop cancer or to have their cancer progress quickly; these behaviours include passive acceptance and self-sacrifice.

Type C personality A behaviour pattern characterized by cooperativeness, unassertiveness, patience, suppression of negative emotions, and acceptance of external authority; supposedly associated with an increased likelihood of cancer.

Type I error The error that is committed when a true null hypothesis is rejected erroneously. The probability of a Type I Error is abbreviated with the lowercase Greek letter alpha.

Type II error The error that is committed when a false null hypothesis is accepted erroneously. The probability of a Type II Error is abbreviated with the uppercase Greek letter beta.

Tyrannophobia Fear of tyrants.

U

Ultimate cause The remotest cause in a chain of causation; e,g., although a person may have died from a stroke, the ultimate cause might have been his fifty previous years of cigarette smoking, which led to the stroke.

Ultradian rhythm The regular recurrence in cycles of less than 24 hours from one stated point to another, as certain biologic activities which occur at such intervals, regardless of conditions of illumination.

Ultrasound examination (sonogram) An examination that involves bouncing high frequency sound waves off the foetus and transforming the bounced waves into visual images.

Ultraviolet light Light whose wavelength is too short to lie within the visible spectrum.

Umbilical cord A flexible, tubelike cord which connects the foetus to the placenta and through which blood, nutrients, oxygen and waste products are interchanged between the mother and the foetus.

Unbiased sample A sample in which every member of the population has an equal chance of being included.

Unconditional positive regard A therapist quality that is considered crucial in client-cantered therapy. It involves nonjudgmental acceptance of the client.

Unconditional response In classical conditioning, a response, such as salivation, that is naturally elicited by the UCS.

Unconditional stimulus (UCS) In classical conditioning, a stimulus, such as food, that naturally elicits a reflexive response, such as salivation.

Unconditioned response (UCR) In classical conditioning, the response elicited by an unconditioned stimulus without prior training or learning.

Unconditioned stimulus (US) In classical conditioning, the stimulus that elicits the unconditioned response (UR) and the presentation of which acts as reinforcement.

Unconscious At a simple and informal level, the notion of an unconscious mind (or subconscious) would seem a usefully straightforward way of accounting for aspects of the mind of which we are not directly conscious or aware. Upon deeper examination, however, the topic reveals extraordinary complexity.

Unconscious inference A process postulated by Hermann von Helmholtz to explain certain perceptual phenomena such as size constancy. An object is perceived to be in the distance and is therefore unconsciously judged to be larger than warranted by its retinal image.

Uncorrelated Variables Lack of relationship between two variables.

Underextension A type of error incurred by a child when it is striving to use a word without comprehending its full adult meaning, for example by using the generic term 'dog' to refer to the family dog.

Underlying pathology model An approach to psychopathology which asserts that various overt signs and symptoms are produced by an underlying cause that may be mental or organic or both. The therapist's objective is to discover and remove the underlying pathology, which will then cause the symptoms to disappear.

Undifferentiated schizophrenia A type of schizophrenia characterized by fragments of the symptoms of other, different types of schizophrenia.

Undifferentiated type A subtype of schizophrenia diagnosed if a patient does not meet criteria for paranoid, disorganized, or catatonic subtypes of schizophrenia.

Undoing Offender whom has a close association with their victim with symbolically try to undo the crime (i.e. wash the victim, place a pillow under their head).

Unemployment The state of being unemployed; lack of employment.

Unipolar depression One of two types of severe depression, where the person only has depressive symptoms without the manic phase.

Unique blue The hue corresponding to a wavelength of 445 nanometers, which is perceived as containing no red or green.

Unique green The hue corresponding to a wavelength of 500 nanometers, which is perceived as containing no blue or yellow.

Unique yellow The hue corresponding to a wavelength of 570 nanometers, which is perceived as containing no red or green.

Universal adaptability In linguistics, the point (believed to be before the age of one) when an infant can distinguish sounds from any language and reproduce them. This flexibility is lost after the child begins to specialize in his or her native language.

Universal grammar (UG) The theory, suggested by the American linguist Noam Chomsky, that all children possess an in-built faculty for spoken language, a theory which is disputed by others, who consider that children are born only with general learning processes which enable them to learn language amongst other things.

Universal structure The specific syntactic organization that some observers believe is preprogrammed in human language.

Universality thesis The hypothesis, originated by Darwin, that facial expressions worldwide look identical, are perceived identically, and express identical emotions.

Unreinforced trial In classical conditioning, a trial in which the conditioned stimulus (CS) is presented without the unconditioned stimulus (US).

Unstable attribution An inference that an event or behaviour is due to unstable,.

Untracking A focus on having students in mixed-ability groups and holding them to high standards but providing many way to reach those standards.

Upper confidence level The upper limit of a confidence interval. If prediction states that the true score falls between 80 and 90, then the upper confidence level is 90.

Uranophobia or Ouranophobia Fear of heaven.

Urophobia Fear of urine or urinating.

Uterus A hollow, muscular organ of female mammals in which the ovum is deposited and the embryo and foetus are developed; womb.

V

V1 The primary visual cortex of a primate.

Vaccinophobia Fear of vaccination.

Valid Well-grounded on principles or evidence; able to withstand criticism or objection, as an argument; sound; in testing, describes a test that measures what it purports to measure.

Validity A measure of the degree to which a test is appropriate for its intended use.

Validity coefficient The correlation between test scores and an independent measure of the trait (for example, scores on a test of "college potential" and the gpas the test takers earn when they get to college). Such a correlation is obtained in order to evaluate the extent to which a test measures the intended attribute or trait.

Validity The extent to which a test measures what it was intended to measure.

Values Enduring beliefs about important life goals that transcend specific situations.

Variability The degree to which a distribution of scores vary around the mean. High variability means scores are spread wider apart and low variability means scores are relatively close together. Typical ways of determining variability are the range, interquartile range, semi-interquartile range, variance, and standard deviation.

Variable A trait or characteristic with two or more categories that can be measured—e.g. "gender".

Variable errors An error caused by random differences in experimental conditions, such as the subject's mood or changes in the environment.

Variable interval schedule A schedule of reinforcement similar to a fixed-interval schedule but characterized by a variable time requirement with a particular mean.

Variable ratio schedule A schedule of reinforcement in which a reinforcer is delivered for the first response made after a variable number of responses whose average is predetermined.

Variance (V) A measure of the variability of a frequency distribution. It is computed by finding the difference between each score and the mean (M), squaring the result, adding all these squared deviations, and dividing the sum by the number of cases. If N is the number of scores, then V = sum of $(\text{score} - M)^2 / N$.

Variation The differences found across individuals of any given species in terms of their genetic, biological (size, strength, physiology), and psychological characteristics (intelligence, sociability, behaviour).

Vasoconstriction The constriction of blood vessels brought on by activation of the sympathetic division of the autonomic nervous system. Vasoconstriction occurs in emergencies, when blood is diverted from the skin and internal organs to the muscles. It is also crucial in mammalian thermoregulation; in response to excessive cold, blood is diverted from the skin to reduce heat loss.

Vasodilatation The dilating of blood vessels brought on by activation of the parasympathetic division of the autonomic nervous system. Vasodilatation is one component of mammalian thermoregulation; in response to excessive heat, warm blood flows to the body's surface and results in heat loss by radiation.

Vasopressin (antidiuretic hormone) A hormone manufactured in the hypothalamus and secreted by the pituitary that elevates blood pressure by producing vasoconstriction and instructs the kidneys to conserve water instead of excreting it.

Vedas Ancient Aryan scriptures - Rig, Yajur, Sama and Atharva (1500 to 1000 B.C.).

Vegetative signs Physical manifestations that often accompany major depression, such as loss of appetite and weight loss, weakness, fatigue, poor bowel functioning, sleep disorders (most often early-morning awakenings), and loss of interest in sex.

Ventral tegmental area (VTA) A region in the midbrain containing dopamine-releasing pathways thought to be involved in reward.

Ventricle One of four large, tubular cavities in the brain, lying under the cortex, that are a leftover characteristic of the brain's early, embryonic development.

Venustraphobia Fear of beautiful women.

Verbal channel In an interview situation, the spoken element, as opposed to the various other non-verbal channels of communication used.

Verbal learning Learning of words or facts under various conditions.

Verbophobia Fear of words.

Verminophobia Fear of germs.

Vertebra One of the bones that encase the spinal cord and constitute the vertebral column.

Vesicles The tiny sacs in the presynaptic neuron that contain neurotransmitters.

Vestibular apparatus The receptive organs of the inner ear that contribute to balance and perception of head movement.

Vestibular sac One of a set of two receptor organs in the inner ear that detect changes in the tilt of the head.

Vestibular senses A set of receptors that provide information about the orientation and movements of the head, located in the semicircular canals and the vestibular sacs of the inner ear.

Vestibular system The sensory system involved in balance.

Vestibules Bony cavities on either side of the head that contain the structures of the inner ear.

Vestiphobia Fear of clothing.

Viable foetus A foetus that is able to live outside of the uterus.

Vicarious distress The distress produced by witnessing another's suffering. Like empathic concern, it motivates one to help others, but is less reliable.

Vicarious learning Learning from observation the consequences of others behaviour.

Vicarious reinforcement According to social learning theorists, a form of reinforcement that occurs when someone watches a model being rewarded or punished.

Victimology The complete history of the victim (i.e. personality, lifestyle, traits etc.).

Videodisc Interactive programs that include videos. films. still pictures, and music.

Villi Any of numerous hairlike or fingerlike vascular processes on certain mucous membranes of the body, as of the small intestine, serving to secrete mucus, absorb fats, and so on; or of the chorion in the mammalian placenta, serving in the exchange of food materials, and so on between the mother and the foetus.

Virginitiphobia Fear of rape.

Viruses Noncellular, microscopic particles that replicate themselves within invaded cells.

Visible spectrum The range of wavelengths to which our visual system can respond, extending from about 400 (the wavelength of the colour violet) to 750 nanometers (the wavelength of the colour reddish orange).

Vision Loss Degree of uncorrectable inability to see well.

Visual agnosia The inability of a person, who is not blind, to recognize the identity or use of an object by means of vision; usually caused by damage to the brain.

Visual cliff A device for assessing depth perception, often used with human infants; it consists of a glass surface that extends from a shallow side over an apparently deep side (the cliff).

Visual cortex Region of the brain's cerebral cortex that processes visual information.

Visual illusion A mental distortion of perception, sometimes misleadingly called an optical illusion, where visual stimuli are seen incorrectly.

Visual pigments Light-sensitive chemicals within the rods and cones of the eye.

Visual search task A test in which research participants are briefly presented a display and must indicate whether a certain target is present or absent.

Vitricophobia Fear of step-father.

Voiced stop A type of phoneme. Main voiced stops as /b/, /d/ and /g/.

Voice-onset time The delay between the initial sound of a voiced consonant (such as the puffing sound of the phoneme /p/) and the onset of vibration of the vocal cords.

Voir dire The legal term used to refer to jury selection. During this process the judge and the attorneys can dismiss prospective jurors for both specific and unstated reasons. Attorneys often examine the prospective jurors age, intelligence, gender, attentiveness, occupation, and open-mindedness.

Volition The motivation or will to make something happen, to reach one's goal.

Volley principle An extension of frequency theory which proposes that when peaks in a sound wave come too frequently for a single neuron to fire at each peak, several neurons fire as a group at the frequency of the stimulus tone.

Volume perception The understanding that containers of different shapes or proportions may hold the same volume.

Volume receptors Receptors that help regulate thirst by indicating the total volume of fluids in the body.

Volunteer bias An error in research that occurs when a sample of volunteers is not representative of the general population.

Vomeronasal organ A distinct set of receptor cells in the nose that are specialized for the detection of pheromones.

Vulnerability Increased likelihood of a particular outcome.

Vulnerability-stress model The idea that individuals who have a biological vulnerability to a particular disorder will have the disorder only if certain environmental stressors are present.

W

Wait time Length of time that a teacher allows a student to take to answer a question. Calling order The order in which students are called by the teacher to answer questions asked during the course of a lesson.

Walloonphobia Fear of the Walloons.

Wason selection task An abstract version of the selection task where the subject is shown one face of each of four cards, but not the other face, and has to select those cards that must be turned over to find out if a particular rule is true.

Wavelength In a periodic waveform, the distance between any point on one cycle and the equivalent point on the next cycle.

Weapon focus effect The tendency for the presence of a weapon to draw attention and impair a witness's ability to identify the culprit.

Weapons effect The tendency of weapons to increase the likelihood of aggression by their mere presence.

Weber fraction In Weber's law, the fraction given by the change in stimulus intensity (delta) divided by the standard intensity (*I*) required to produce a just-noticeable difference: *I/I* = *C*.

Weber's law The amount of change necessary to detect a different in a stimuli must be staged in a percentage since recognition of the change is relative to the characteristics of the initial stimulus (e.g., a one pound change in the weight of a pencil would be more easily recognized than a one pound weight added to a 300 pound barbell because it represents a much greater percentage of the total weight).

Wechsler adult intelligence scale (WAIS) WAIS is a general test of intelligence (IQ), published in February 1955 as a revision of the Wechsler-Bellevue test (1939), standardised for use with adults over the age of 16. Intelligence is quantified as the global capacity of the individual to act purposefully, to think rationally, and to deal effectively with his environment. The full scale IQ is broken down into 14 subtests, comprising the verbal (7 subtests) and performance scales (7 subtests).

Wechsler intelligence scale for children (WISC) An intelligence test for children devised by David Wechsler; similar in form to the Wechsler Adult Intelligence Scale.

Well-defined problems Problems for which there is a clear-cut way of deciding whether a proposed solution is correct. This contrasts with ill-defined problems, for which it is unclear what a correct solution might be.

Wellness Optimal health, incorporating the ability to function fully and actively over the physical, intellectual, emotional, spiritual, social, and environmental domains of health.

Wernicke's aphasia A condition, described by German neurologist Carl Wernicke, where an inability to understand what words mean affects a persons ability to produce meaningful utterances.

Wernicke's area A region of auditory association cortex located in the upper part of the left temporal lobe; involved in the recognition of spoken words.

What is beautiful is good stereotype The belief that physically attractive individuals also possess desirable personality characteristics.

White matter The portions of the central nervous system that are abundant in axons rather than cell bodies of neurons. The colour derives from the presence of the axons' myelin sheaths.

Whole class discussion A discussion among all the students in a class with the teacher as moderator.

Whole language An educational philosophy that emphasizes the integration of reading, writing, and language and communication skills across the curriculum in the context of authentic or real-life materials, problems, and tasks.

Whole-word reading Reading by recognizing a word as a whole; "sight reading.".

Wiccaphobia Fear of witches and witchcraft.

Widow A woman who has outlived the man to whom she was married at the time of his death; especially, such a woman who has not remarried.

Widower A man who has outlived the woman to whom he was married at the time of her death; especially, such a man who has not remarried.

Widowhood The disruption of marriage due to the death of the spouse.

Wisdom As defined by Erik Eriksson, this is the quality that results from the resolution of ego integrity and despair, which is achieved by the adult in old age, when he/she has been able to find a midway between the calm acceptance of a life lived and the fear of impending death.

Wit work The process by which a denied impulse (invariably hostile or obscene) is converted into a joke. Similar to dreamwork.

Withdrawal One of the 'ethical principles for conducting research with human participants' endorsed by the British psychology society in which a subject retains the right to his/her 'data' and is able to withdraw consent at any time and have any data pertaining to them excluded from the study.

Withdrawal symptoms A consequence of drug addiction that occurs when the drug is withheld. These effects tend to be the opposite of those produced by the drug itself.

Within class ability grouping A system of accommodating student differences by dividing a class of students into two or more ability groups for instruction in certain subjects.

Within-family differences In research on the genetics of behaviour, a term often used to refer to the role of environment. It describes differences in the environment (e.g., different schools) of individual family members. For most personality attributes, these seem to be more important than between-family differences.

Within groups Design A design that compares a group's results to a measure taken before treatment.

Within-group heritability The extent to which variation within groups (e.g., among U.S. whites) is attributable to genetic factors.

Within-subjects design A research design that uses each participant as his or her own control; for example, the behaviour of an experimental participant before receiving treatment might be compared to his or her behaviour after receiving treatment.

Withitness The degree to which the teacher is aware of and responsive to student performance.

Womb envy In Karen Horney's view, the discontent and resentment that men experience because of their inability to bear children.

Women Needier, more demanding of affection, less governed by conscience, more prone to neurosis and hysteria, fundamentally passive, more narcissistic. (In all fairness, it should be noted that Freud, a strong advocate for women becoming analysts, worked actively and vocally against attempts to block women from entering his profession.).

Word processing A computer application for writing compositions that lends itself to revising and editing.

Workaholism Addiction to work.

Working memory A memory resource that is used to accomplish tasks such as reasoning and language comprehension; consists of the phonological loop, visuospatial sketchpad, and central executive.

Working self-concept A temporary structure in memory which is invoked by the requirements of a particular situation and forms the configuration of self-views made salient by ongoing social events.

Worry Something or someone that causes anxiety; a source of unhappiness.

X

Xanthophobia Fear of the colour yellow or the word yellow.

Xenoglossophobia Fear of foreign languages.

Xenophobia Fear of strangers or foreigners.

Xerophobia Fear of dryness.

Xylophobia .1 Fear of wooden objects.
2. Forests.

Xyrophobia Fear of razors.

Y

Y chromosome The Y chromosome is one of the two sex chromosomes in humans and most other mammals (the other is the X chromosome). The sex chromosomes are one of the 23 pairs of human chromosomes. The Y chromosome spans about 50 million base pairs (the building blocks of DNA) and represents between 0.5 and 1 percent of the total DNA in cells. The chromosome may be used to trace parental lineage.

Yerkes Dodson law Theory arguing that for performance to be optimal, the amount of arousal required must be optimized. Too much or too little stimulation will result in a poorer performance.

Yogabhrashta False or pseudo yogi, disqualified from the yogic path because of some error in intention, life-style or practice.

Young adulthood That period of human development, as defined by Erik Eriksson, when a person develops the need and capacity for intimacy and the ability to be committed to other people despite loss of attention to their own needs.

Z

Z score The conversion of a score to a standard deviation. This number tells how many standard deviations a score is from the mean.

Zeitgeber Any stimulus, such as light, that synchronizes daily biological rhythms.

Zeitgeist Zeitgeist is originally a German expression, which means "the spirit (Geist) of the time (Zeit)". It denotes the intellectual and cultural climate of an era.

Zelophobia Fear of jealousy.

Zemmiphobia Fear of the great mole rat.

Zero correlation the absence of a relationship between two or more variables as determined by a correlational statistic. Often abbreviated as 'r=0.'.

Zeusophobia Fear of God or gods.

Zona pellucida The gelatinous covering of the egg.

Zone of proximal development Those aspects of a child's performance, described by the Russian psychologist Lev Vygotsky, where she would not be able to achieve on her own, but only with the support, help and instruction of another person.

Zoophobia Fear of animals.

Zygote A cell that results from the combination of a sperm cell and an egg during.

Zygote stage The first stage of prenatal development, during which the zygote divides many times and the internal organs begin to form.